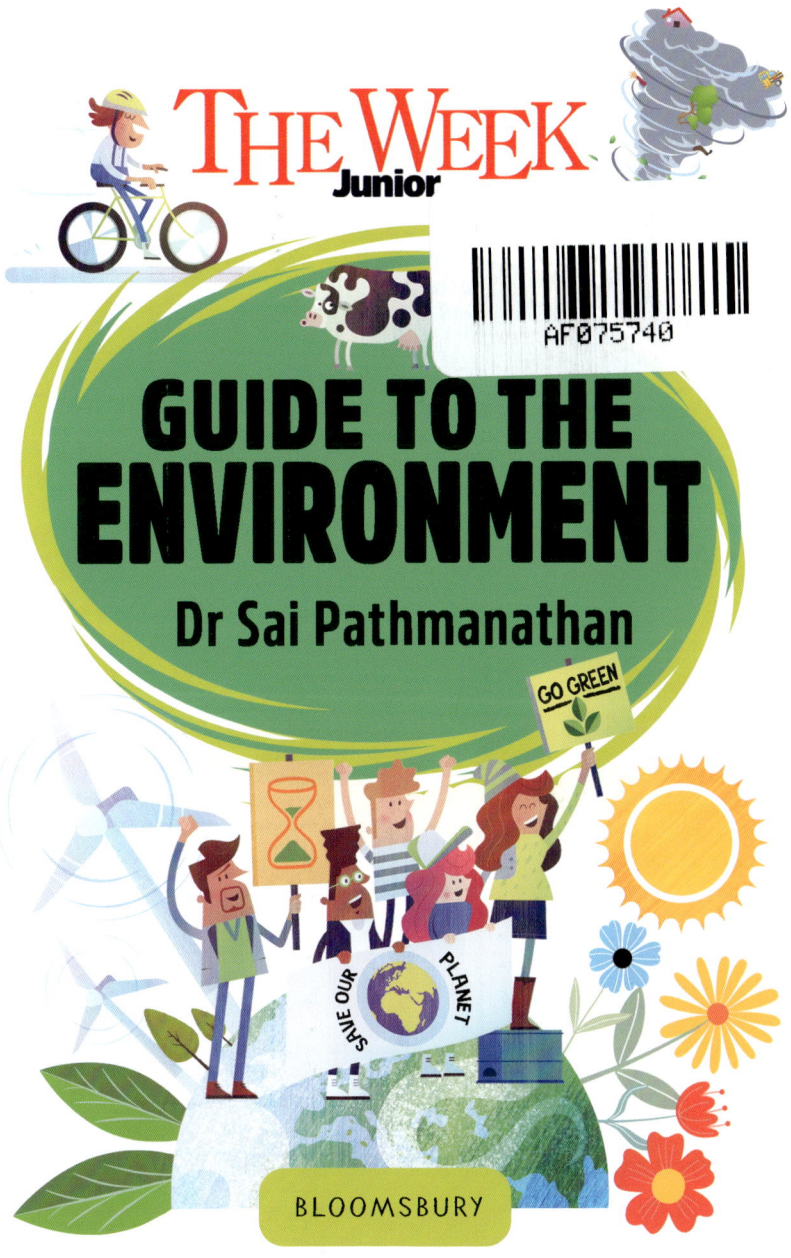

BLOOMSBURY EDUCATION
Bloomsbury Publishing Plc
50 Bedford Square, London, WC1B 3DP, UK
Bloomsbury Publishing Ireland Limited
29 Earlsfort Terrace, Dublin 2, D02 AY28, Ireland

BLOOMSBURY, BLOOMSBURY EDUCATION and the Diana logo are trademarks of Bloomsbury Publishing Plc
First published in Great Britain, 2025 by Bloomsbury Publishing Plc

Text copyright © Sai Pathmanathan, 2025
Illustrations copyright © Tim Bradford, 2025

The Week Junior and The Week Junior logo © Future Publishing Limited, 2025

Sai Pathmanathan and Tim Bradford have asserted their rights under the Copyright, Designs and Patents Act, 1988, to be identified as Author and Illustrator of this work

Every reasonable effort has been made to trace and acknowledge copyright holders of material reproduced in this book, but if any have been inadvertently overlooked the publishers would be glad to hear from them. For legal purposes, the picture credits on p122 constitute an extension of this copyright page. Some images in this book were created to illustrate, are artists' impressions, or are reconstructions.

Bloomsbury Publishing Plc does not have any control over, or responsibility for, any third-party websites referred to or in this book. All internet addresses given in this book were correct at the time of going to press. The author and publisher regret any inconvenience caused if addresses have changed or sites have ceased to exist, but can accept no responsibility for any such changes.

All rights reserved. No part of this publication may be: i) reproduced or transmitted in any form, electronic or mechanical, including photocopying, recording or by means of any information storage or retrieval system without prior permission in writing from the publishers; or ii) used or reproduced in any way for the training, development or operation of artificial intelligence (AI) technologies, including generative AI technologies. The rights holders expressly reserve this publication from the text and data mining exception as per Article 4(3) of the Digital Single Market Directive (EU) 2019/790

A catalogue record for this book is available from the British Library

ISBN: PB: 9781801995856; ePDF: 9781801995870; ePub: 9781801995863
2 4 6 8 10 9 7 5 3 1

Text design by Sophie Gordon and Laura Neate

Printed and bound in India by Manipal Technologies Limited

To find out more about our authors and books visit www.bloomsbury.com and sign up for our newsletters

For product safety related questions contact productsafety@bloomsbury.com

THE WEEK Junior

GUIDE TO THE ENVIRONMENT

Dr Sai Pathmanathan
Illustrated by Tim Bradford

BLOOMSBURY EDUCATION
LONDON OXFORD NEW YORK NEW DELHI SYDNEY

ADVERTISEMENT

Fuel Fascination
with

The Week Junior is a **multi-award-winning children's magazine** packed with engaging articles, eye-catching images, and big ideas that get eight to 14-year-olds reading, thinking, and talking.

Published weekly, it features everything from news to sport, science, books, cooking, and craft – **feeding children's natural curiosity, encouraging critical literacy and boosting confidence.**

Get 6 issues for £1 at
theweekjunior.co.uk/offer

OFFER CODE **PBLOOM25**

CONTENTS

How To Use This Book		vi
Introduction		1
Chapter 1	**Climate Change**	8
Chapter 2	**Weather and Natural Disasters**	26
Chapter 3	**Polluting Our Planet**	36
Chapter 4	**The Variety of Life on Earth**	46
Chapter 5	**One Health**	60
Chapter 6	**Food and Farming**	68
Chapter 7	**Leisure and Entertainment**	78
Chapter 8	**Fast Fashion**	86
Chapter 9	**Travelling Around**	96
Chapter 10	**Money and Power**	108
Find Out More		118
Glossary		119
Index		121

HOW TO USE THIS BOOK

An explanation of a key concept

A brief introduction to an important person

A key word defined in the glossary at the back of the book

A fun fact

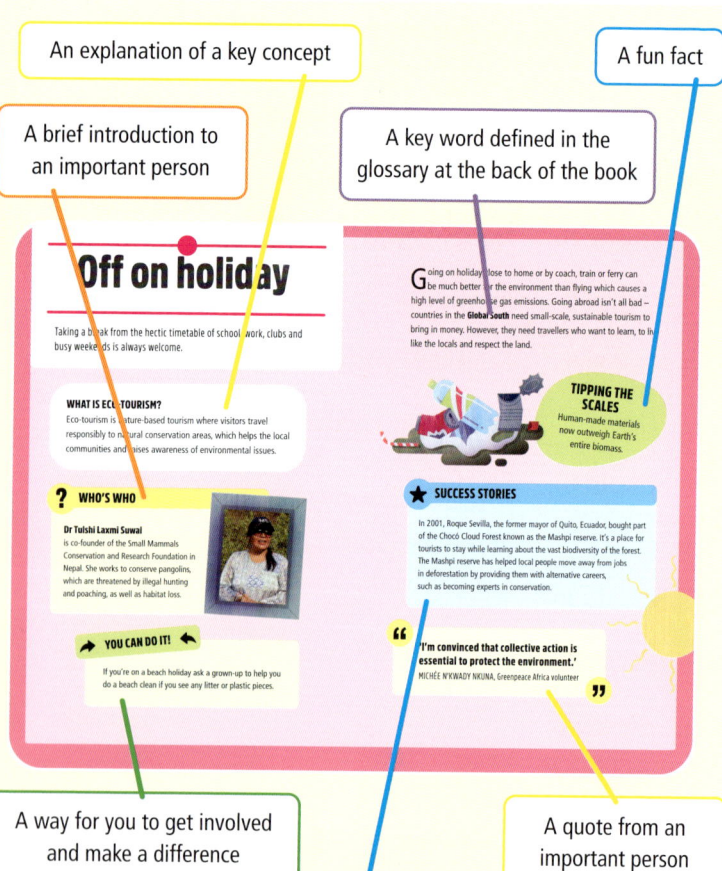

Off on holiday

Taking a break from the hectic timetable of school, work, clubs and busy weekends is always welcome.

Going on holiday close to home or by coach, train or ferry can be much better for the environment than flying which causes a high level of greenhouse gas emissions. Going abroad isn't all bad – countries in the **Global South** need small-scale, sustainable tourism to bring in money. However, they need travellers who want to learn, to live like the locals and respect the land.

WHAT IS ECO-TOURISM?
Eco-tourism is nature-based tourism where visitors travel responsibly to natural conservation areas, which helps the local communities and raises awareness of environmental issues.

TIPPING THE SCALES
Human-made materials now outweigh Earth's entire biomass.

? WHO'S WHO
Dr Tulshi Laxmi Suwal is co-founder of the Small Mammals Conservation and Research Foundation in Nepal. She works to conserve pangolins, which are threatened by illegal hunting and poaching, as well as habitat loss.

★ SUCCESS STORIES
In 2001, Roque Sevilla, the former mayor of Quito, Ecuador, bought part of the Chocó Cloud Forest known as the Mashpi reserve. It's a place for tourists to stay while learning about the vast biodiversity of the forest. The Mashpi reserve has helped local people move away from jobs in deforestation by providing them with alternative careers, such as becoming experts in conservation.

➤ YOU CAN DO IT! ◄
If you're on a beach holiday ask a grown-up to help you do a beach clean if you see any litter or plastic pieces.

" **I'm convinced that collective action is essential to protect the environment.** "
MICHÉE N'KWADY NKUNA, Greenpeace Africa volunteer

A way for you to get involved and make a difference

An environmental win

A quote from an important person

INTRODUCTION

Our planet is incredible; it gives us so much. The air we breathe, the magnificent trees, unique plants and amazing wildlife: from the tiniest tree frogs in our rainforests to the huge whales in our oceans. However, all isn't well on planet Earth. You've probably heard people talking about things like saving the environment, the climate crisis, pollution, **microplastics** and **greenhouse gases**. It can feel like there's a lot of pressure to solve the planet's problems. However, many inspiring people and organisations are working together to save our Earth – and you can help too.

> '**I'm telling you there is hope […] it does not come from the governments or corporations. It comes from the people.**'
> GRETA THUNBERG, climate justice activist

What's the problem?

The Earth has existed for millions of years and has seen **asteroids**, dinosaurs and **ice ages**. So what's the issue now?

If we imagine Earth as 46 years old rather than around 4,600 million years old, then its life story goes like this…

1–42 years old: We know hardly anything about this time

42–45 years old: Living **organisms evolve**

Age 45 (last year): Dinosaurs rule

Humans have only been around for a small part of the life of our planet, but we have had a big impact on it. The way we produce and use food, homes, transport, clothes and more is using up the Earth's resources. We've destroyed **habitats** animals need to live. We've been burning **fossil fuels**, causing the world to get warmer and creating problems for wildlife and humans across the globe. We've been throwing lots of stuff away, and the waste we create is causing problems for the planet.

The reality of problems like **climate change** and plastic pollution is quickly sinking in, and we need to take action now to protect ourselves and the planet.

8 months ago: Mammals evolve

Age 46 (this year):
- This weekend: The Ice Age (brrr)
- 4 hours ago: Humans evolve
- 1 hour ago: Humans discover agriculture
- 1 minute ago: The **Industrial Revolution**
- The past few seconds: Humans turn the planet into a rubbish tip

What can you do?

First of all, don't panic. Climate change and the other environmental problems sound very worrying and the news reports can be scary. However, although the problems are urgent, there's a lot we can do together. The fight against climate change is happening right now.

Each of our individual actions can help, like walking rather than driving, going on holiday by train rather than plane, switching off lights we don't need, eating less meat and buying less single-use plastic. We also need to put pressure on governments and world leaders to make big changes.

Many young people are joining the climate campaign and protesting for positive change – maybe you've joined a protest yourself. Protest marches such as the school strikes for climate, known as Fridays For Future, get noticed by those who can change laws. Decision makers actually see how many people want change. Find what you're passionate about and use your talents to inspire others and bring about change. Making the planet better for everyone can start with you.

GLOBAL CLIMATE STRIKE
The world's largest climate protest took place in 2019, involving over four million people across 163 countries.

YOU CAN DO IT!

If there's a protest happening near where you live, ask an adult if you can go and make your voice heard. Make bold signs with your message on it, shout and use musical instruments to grab attention. But remember, the most impactful protests are ones carried out peacefully, stating the facts and making your concerns heard. Make sure you stay in a group, carry plenty of water and snacks and wear comfortable, weather-suitable clothing.

Power of the people

Not all of us feel we can be the next **activist** or confident public speaker like Greta Thunberg. We can, however, all do something if we care about our planet.

Activism isn't just about protests. There are lots of ways to make your voice heard. It could be writing to a local leader, persuading them to make changes. It could be taking impactful photos of plastic pollution or local wildlife to raise awareness. What about writing poems or creating posters to display at school or a local community venue so everyone talks about **biodiversity**? Maybe you could raise funds or volunteer for a local environmental charity.

Taking photos can raise awareness

Protesters in Australia

Climate change and other environmental issues affect all parts of our lives. They are complicated problems to solve and saving the planet is a team effort. In this book, you'll find the stories of many inspiring people and organisations who are changing things for the planet, as well as ideas for you too. You will see how connected our lives are with the world around us: what we wear, the foods we eat, the products we buy and the places we travel to all make a difference. There are many factors at play and we can do our best to change what we can and hope that others will change with us.

> **'I'm convinced that collective action is essential to protect the environment.'**
> MICHÉE N'KWADY NKUNA, Greenpeace Africa volunteer

CHAPTER 1
CLIMATE CHANGE

WHAT IS CLIMATE?

Climate is the usual weather conditions. For example, the Caribbean has a hot climate whereas countries such as Iceland and Greenland have a very cold climate. Scientists study the weather patterns over a period of over 30 years to say what a region's climate is. A few days of cold and rainy *weather* in a hot and sunny region doesn't mean the *climate* has changed.

The Earth's climate has always changed, long before humans were around. However, scientists have found that global temperatures are rising higher and faster than ever before. Since 1880, the Earth's average temperature has gone up by 1°C, and the last 20 years have seen some of the warmest years on record.

Find out more: greenhouse gases pages 12-13, natural disasters pages 26-27

Wildfires in Alaska, USA

Wildfire damage in Athens, Greece

A dried up riverbed, USA

Flooding in Baco, Philippines

Maybe you're thinking that sunshine all year round sounds pretty good. However, the planet warming is actually causing all sorts of damage. Ice that has been in place for hundreds of years, such as Antarctic sea ice, is shrinking fast. The world is experiencing more frequent **wildfires** (in places like the Mediterranean and the West Coast of the US). Some regions have heavy rain and **floods** while others have **droughts**. These are not one-off events; they are signs of a rapidly warming world. We can no longer assume seasons and climate will be the same in the future.

LET IT HAIL
Christmas in Australia is normally sunny but, in 2023, hailstones fell which were big enough to break windows. Scientists say climate change caused larger hailstones to form.

The greenhouse effect

Greenhouse gases in the Earth's atmosphere trap heat and stop it from escaping. This keeps the Earth warm, just like plants in a greenhouse. Greenhouse gases cause the **greenhouse effect**. Without this effect, the Earth's average temperature would be -18°C.

Throughout most of history, humans didn't create many greenhouse gases. People used candles and oil for lighting and wood for heating, and lived with what they grew and made themselves. This all changed during the Industrial Revolution (1760–1840).

Humans came up with inventions to save time and make their lives easier. Factories with machines, engines and electricity relied on fossil fuels like coal and gas. Other inventions and discoveries such as providing clean water and medicines meant people started living longer. More humans were born, so more homes, cities, roads, farms and factories were needed. All of that produced a lot of greenhouse gases, meaning that less heat could escape into space. We are now seeing an Enhanced Greenhouse Effect, or **global warming**, causing the Earth to heat up. We can't reverse this but we can limit our global temperature increase.

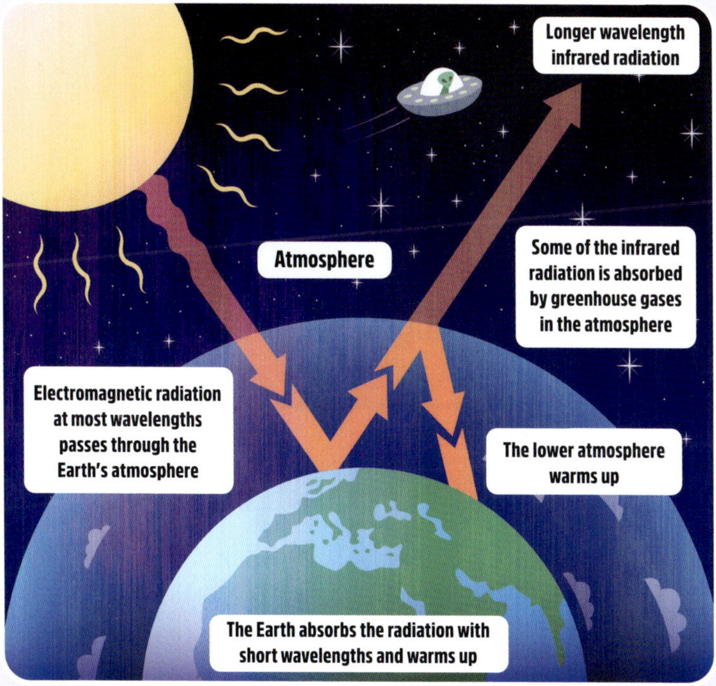

- **Longer wavelength infrared radiation**
- **Atmosphere**
- **Some of the infrared radiation is absorbed by greenhouse gases in the atmosphere**
- **Electromagnetic radiation at most wavelengths passes through the Earth's atmosphere**
- **The lower atmosphere warms up**
- **The Earth absorbs the radiation with short wavelengths and warms up**

➤ YOU CAN DO IT! ⬅

Imagine you are planet Earth. Put a jumper on. The jumper acts like the greenhouse effect. Now add another jumper or a coat. How warm do you feel? Add a blanket.
You'll start to feel hot and sweaty.
This is the Enhanced Greenhouse Effect.
The extra layers are like the greenhouse gases in the Earth's atmosphere, trapping more heat.

Find out more: carbon emissions pages 14-15, global warming pages 8-9

What are the main greenhouse gases?

Water (H_2O)
Water vapour in the atmosphere is natural; it's part of the **water cycle** and important in the normal greenhouse effect. As the Earth warms further, more liquid water **evaporates** into water vapour so there is more in the atmosphere.

Carbon dioxide (CO_2)
Billions of **tonnes** of **carbon dioxide** go into the atmosphere from activities like burning wood, coal and other fossil fuels, **deforestation** and making cement.

❓ WHO'S WHO

Eunice Foote (1819–1888) was an American scientist who was the first to describe how carbon dioxide contributes to climate change. She put thermometers into two glass cylinders: one containing air, the other carbon dioxide. She measured the temperatures after they had been in the sun for a while and the one containing carbon dioxide was hotter. She worked out that this meant carbon dioxide would absorb heat from the atmosphere.

Nitrous oxide (N_2O)

This gas is released into the atmosphere by farming, particularly the use of chemical **fertilisers**, deforestation, burning fossil fuels and driving vehicles.

Methane (CH_4)

This is released into the atmosphere through burning and extracting fossil fuels, farming animals and **landfill sites**. One tonne of methane can trap the same amount of heat as 28 tonnes of carbon dioxide.

Other gases

These include hydrofluorocarbons used in fridges and air conditioning systems, and sulphur hexafluoride used to insulate power lines.

Carbon jargon

Scientists and environmentalists use the word 'carbon' a lot. Sometimes they're talking about the greenhouse gas carbon dioxide; 'carbon' can also refer to greenhouse gases more generally. Often the symbol CO_2e is used for 'carbon dioxide equivalents' which include other greenhouse gases. Here are some other terms that are useful to know.

CARBON FOOTPRINT
The total amount of greenhouse gases produced by human activity.

CARBON OFFSETTING
Increasing carbon storage (by planting trees, for example) to make up for greenhouse gas **emissions** that happen somewhere else (like a plane flight).

STORED CARBON
Carbon locked away in rocks and soil, dissolved in the oceans and stored in living things such as plants. The more plants we have on Earth the more carbon is stored.

NEW TREES
In 2019, over 23 million Ethiopians took part in a tree-planting day and planted over two hundred million new trees.

CARBON-NEUTRAL
Balanced between emitting and absorbing carbon.

CARBON CAPTURE
Removing and reducing carbon dioxide in the air. This may be achieved by planting trees (trees absorb carbon dioxide and release oxygen) or using special technology allowing carbon dioxide to be stored underground or combined with other materials to make rocks, bricks or carbon fibre.

DECARBONISATION
Reduction or elimination of carbon emissions.

Find out more: carbon capture pages 22-23

Fossil fuels

Many of the greenhouse gases causing problems for our climate are caused by burning fossil fuels. Fossil fuels are known as non-renewable, meaning they can't be replaced and will eventually run out.

WHAT ARE FOSSIL FUELS?

Fossil fuels such as coal, oil and natural gas are made from the remains of prehistoric animals and plants. These – in common with all living things – contain carbon. As fossil fuels burn, the carbon reacts with **oxygen** in the air, producing carbon dioxide gas and other waste products.

Most power stations today burn coal, oil or gas to produce electricity. Burning coal can produce more than twice its weight in carbon dioxide emissions; this is why we need **renewable** sources to create electricity instead.

In 2019, Great Britain went seven days without using coal to produce electricity; this was the first time since 1882. In 2020, this had increased to 68 days continuously without using coal. While coal remains one of the main sources of electricity production in the world, the United Kingdom **phased out** coal completely by October 2024.

Getting hold of fossil fuels is also difficult, dangerous, expensive and harmful to local wildlife. **Mining** coal leaves the land unstable and releases toxic fumes. Drilling and **fracking** for oil and gas risks explosions as well as poisonous fumes. Oil spills can kill seabirds, fish and other wildlife.

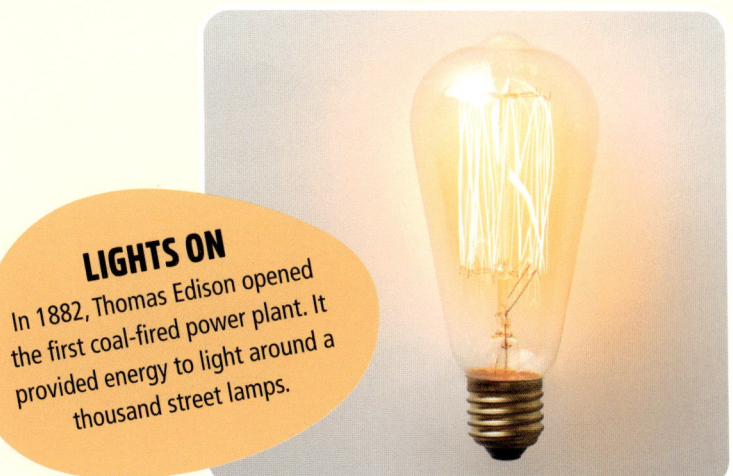

LIGHTS ON
In 1882, Thomas Edison opened the first coal-fired power plant. It provided energy to light around a thousand street lamps.

Find out more: renewable energy pages 22-23, transport emissions pages 96-97

Talks and targets

As it has become clearer that people are having an impact on the climate, scientists have been busy finding ways to reduce our need for fossil fuels. However, some people don't want change. Some think they should still be able to use as many fossil fuels as they want. Some politicians, especially in the **Global North**, think the changes needed could be bad for businesses or unpopular with voters.

Every year, world leaders, environmentalists and scientists meet at the **United Nations** Climate Change Conference, known as the Conference of the Parties (COP) to discuss **policies** for improvement. In 2015, they created a legal document known as the Paris Agreement. In it, 195 countries agreed to make sure the global average temperature increase stays below 2°C. More recently, world leaders have said we need to limit global warming to 1.5°C before the end of this century. The global average temperature had already increased by 1.1°C in 2021 and is increasing by 0.2°C every ten years.

To stop this, we need to reduce our greenhouse gas emissions by at least half by 2030. To get us on track, countries and companies around the world are setting more ambitious targets. The UK aims to reach **net zero** by 2050.

We'll all need to work together to reach net zero

WHY DO RISING TEMPERATURES MATTER?

1.5°C might seem like a small change in temperature but the Intergovernmental Panel on Climate Change reports that if we go above this, severe climate events such as droughts, heatwaves and heavy rainfall will become more likely.

Saving energy

Even with all the work that's been done, a lot of the electricity we use comes from power stations that burn fossil fuels. Whenever we plug in a phone, tablet or kettle, we're creating greenhouse gases. Regularly saving energy is one important way to tackle the climate crisis.

Instead of turning the heating up, put on a jumper or wrap up in a blanket. Lowering the temperature of your heating by just 1°C can stop 340kg of carbon dioxide entering the atmosphere each year.

YOU CAN DO IT!

Make a draught excluder by filling a pair of old tights or socks with scrap fabric or sand. Tie the ends and place it in a doorway, or wherever you feel there's a cold breeze coming in.

When the heating or air conditioning is on, shut the windows to stop air escaping and wasting energy.

Ask your grown-up to check that your boiler is running efficiently by getting a plumber to service it every year.

Check if your light bulbs are energy-efficient ones – there should be a rating on the packaging. The most efficient ones (compact fluorescent and LED bulbs) use 90 per cent less electricity than traditional ones.

Make sure you only boil as much water as you need in the kettle. If you do boil too much, keep it warm in a vacuum flask for later, or use it for a hot water bottle.

Turn off lights and electrical devices when you're not using them (don't leave them on **standby**).

Ask your grown-ups to check how well **insulated** your home is.

Find out more: renewable energy pages 22-23, sustainable homes pages 112-113

Renewables

Renewable or clean energy doesn't get used up or emit greenhouse gases like fossil fuels do. Climate scientists say we need the world's energy supplies to be 100 per cent renewable by 2050 at the latest to tackle the climate crisis.

A few countries such as Iceland and Costa Rica already get most of their electricity from a mix of renewable sources, but other countries have work to do. Renewables are much cheaper than fossil fuels once they're up and running so many countries will see the financial benefit. However, unless we phase out fossil fuels completely, the planet will keep getting warmer.

DIFFERENT TYPES OF RENEWABLE ENERGY

SOLAR

Solar panels made of silicon and glass absorb sunlight which is converted into electricity. The biggest solar farm in the UK is capable of powering 14,000 homes.

Solar panels

PLAYER POWER

A sports pitch in Rio de Janeiro, Brazil, has lights powered by solar panels and by special tiles that turn the players' movements into energy.

WIND

Wind turns **turbines** to generate electricity. Large wind farms can produce enough electricity to power 300,000 homes but when it's not so windy, turbines can't generate enough power.

Wind turbines

WOOD WIND

A company in Gothenburg, Sweden makes turbine towers out of wood. Wood is lighter and easier to form and transport than steel.

A geothermal power station

GEOTHERMAL

Uses heat from deep underground to power **generators**. Almost 90 per cent of homes in Iceland are heated by geothermal energy.

Renewables

It's worth remembering that these sources of renewable energy aren't perfect – they all have some benefits and some risks.

GREEN HYDROGEN

Used to fuel cars, aircrafts and spacecraft and can be used to generate electricity and heat our homes. Hydrogen is produced through a process called electrolysis where electricity is used to split water into hydrogen and oxygen. When this process uses electricity from renewable sources it creates **green** hydrogen. Unfortunately, this is more expensive than creating hydrogen using fossil fuels, which emits 830 million tonnes of carbon dioxide into the atmosphere every year.

A hydrogen engine

WATER (ALSO KNOWN AS HYDROELECTRIC POWER)

Water turns turbines to generate electricity. This can be the downhill flow of a river, or countries with a long coastline can use the energy from waves and tides.

Turbines turned by water flow

BIOMASS

A biomass processing plant

Plant or animal matter which is burnt to generate electricity. It's called renewable as it won't run out, but it's not clean. Burning biomass might keep fossil fuels in the ground, but it can create the same if not more greenhouse gas emissions and cause long-lasting damage to forests.

NUCLEAR POWER

Uses the huge energy stores in atoms of uranium by splitting them apart in a reactor. Nuclear power is a low carbon energy source and is reliable – it provides energy whatever the weather.

A nuclear power plant

However, it also produces **radioactive** waste which is expensive to dispose of. Accidents at nuclear power stations can be dangerous for humans and the wider environment.

CHAPTER 2
WEATHER AND NATURAL DISASTERS

Global warming is changing the weather depending on where we live.

WHAT IS WEATHER?

Humans love talking about the weather, whether it's sunny or pouring with rain. Weather is different from climate; it is simply the conditions outside on a given day or few days. Sunny, raining, snowing and windy are all ways to describe the weather.

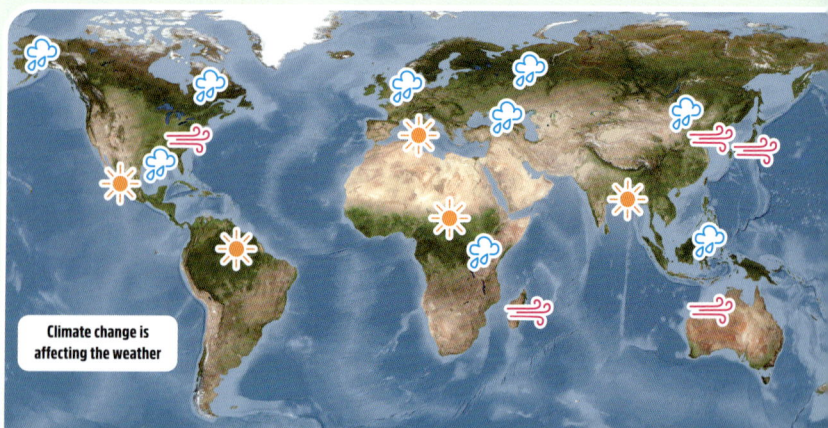

Climate change is affecting the weather

HEAT

We are seeing record-breaking temperatures and severe heatwaves. In deserts and dry places, hotter temperatures cause droughts and wildfires. 2024 was the hottest year on record. If this carries on, we'll lose a lot of wildlife and more trees than we can replant. Heat has an impact on colder regions too – see pages 28-29 for more information.

WIND

Wind is air in motion caused by differences in air pressure. Gases move from high-pressure (cooler air) to low-pressure (warmer air) areas. The bigger the difference in pressure, the faster the air moves. Changing global temperatures can lead to stronger windstorms, but scientists also say that as carbon dioxide levels rise and the Earth's poles get warmer, there may be a slowing of wind speeds. This could be an issue for generating wind power.

RAIN

Scientists say that extreme rainfall is expected to increase as the planet warms. Heat makes more water evaporate from the oceans which falls as rain on land nearby. A lot of rain in a short time leads to more floods, which can damage buildings and spread dirty water into clean water supplies, causing disease.

Find out more: climate change pages 8-9, biodiversity pages 46-47

As the ice melts

About a tenth of the Earth's land is covered in glaciers and ice sheets. Glaciers are huge sheets of slow-moving ice which occur in cold, snowy places such as at the North and South Poles or on high mountains.

As the planet warms, this ice is melting rapidly, which releases water which flows into the oceans. This is causing the sea level to rise. If all the glaciers melt, the sea level will rise by 70 metres.

Melting glaciers are causing sea levels to rise

REST IN PEACE
Okjökull, a glacier in Iceland, has shrunk so much it's become too small to be counted as a glacier. In 2019, the Icelandic president held a funeral for it.

Melting ice also affects all wildlife in cold regions. The Arctic sea ice helps polar bears travel around to hunt for food, so the melting ice makes it harder for them to hunt. Antarctic krill (shrimp-like creatures) feed on the **algae** that grows just under the ice. As the ice melts, krill numbers fall. This affects whales, seabirds, fish, squid, sharks and seals as they all need krill as a source of food.

WHAT IS THE ALBEDO EFFECT?

As ice is pale and bright, it can reflect sunlight back out into space. As the ice melts, the Earth's albedo, or reflectiveness, drops. The less ice there is, the less sunlight is reflected, causing the Earth to warm up more and even more ice to melt.

AMOUNT OF SUNLIGHT REFLECTED BY DIFFERENT SURFACES

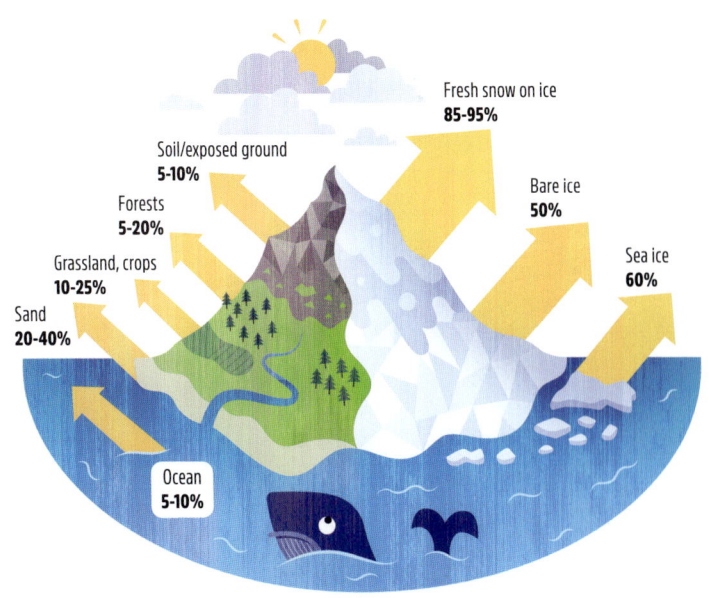

Fresh snow on ice
85-95%

Soil/exposed ground
5-10%

Bare ice
50%

Forests
5-20%

Sea ice
60%

Grassland, crops
10-25%

Sand
20-40%

Ocean
5-10%

Natural disasters

Climate change is causing natural disasters to become more damaging and more common.

EARTHQUAKES AND TSUNAMIS

Droughts caused by climate change can lead to the deterioration of **fault lines** in the **Earth's crust**. Pumping groundwater (the water under the Earth's surface) for our use during droughts can put further pressure on the Earth's crust and result in earthquakes. Earthquakes are the shaking of the ground caused by movements within the Earth's crust. They can lead to tsunamis – huge waves caused by earthquakes – or volcanic eruptions under or near the sea. The earthquake causes the ocean floor to lift or fall, causing a huge volume of water to move.

Earthquake damage in Turkey

VOLCANIC ERUPTIONS

The movement of the **Earth's crust** during earthquakes can cause volcanoes to become active. In 2023, one of Iceland's volcanoes, around 25 miles from the capital Reykjavík, started to erupt, forcing lava more than 100 metres into the sky.

A volcano in Iceland

TYPHOONS, CYCLONES AND HURRICANES

Whirling windstorms have different names depending on where they form. Typhoons form in the western Pacific, cyclones in the Indian Ocean and hurricanes in the Atlantic. Climate change has made the oceans warmer, which causes more water to evaporate which, in turn, leads to stronger storms, faster wind speeds and more destruction.

Hurricane damage in Florida, USA

? WHO'S WHO

Marinel Samook Ubaldo
a young climate activist who lived through Typhoon Haiyan, campaigns for bans on single-use plastics and the reduction of emissions, and has successfully taken the biggest polluters in the Philippines to court.

Find out more: extreme weather pages 26-27, human migration pages 34-35

Adapting to changing weather

As the planet warms up, weather events are changing and **species** are having to adapt to survive.

El Niño is the warming of the ocean surface in the central and eastern tropical Pacific Ocean. Every two to seven years, the warm waters move west and surround the Galápagos Islands. Warmer waters around the islands lead to less algae, which is usually eaten by the iconic marine iguana.

Iguanas are shrinking

Some marine iguanas die of hunger, but others adapt by shrinking. Their skeletons become smaller as they reabsorb their bone matter and they exist as mini versions of themselves, a fifth of their original size. Marine iguanas that shrink are thought to survive longer because they need less food and can forage using less energy. When the water cools and the algae return, the marine iguanas grow to their usual size again. However, if El Niño events become more common with the warming climate, how much will the iguanas have to shrink?

SUCCESS STORIES

When wildlife finds it difficult to adapt to changing conditions, **conservation** scientists can step in to help. For example, in Australia, as carbon dioxide levels have increased, droughts and bushfires have become more common, and koalas are losing their homes and food: eucalyptus trees. Organisations like The World Wide Fund for Nature Australia are trying to help by using drones to spread 40,000 eucalyptus seeds a day.

Humans and the changing weather

Climate change is causing mass human migration.

Forty per cent of the world's population lives within 100 km of the sea; rising sea levels and more powerful storms will force people to move. Droughts, heatwaves, wildfires, earthquakes, tsunamis, eruptions and storms can destroy homes. These disasters seriously threaten many people around the world, especially those in rural and **indigenous** communities who rely heavily on natural resources. We all have to start taking notice. It's happening everywhere.

> **'I don't want to die knowing that while I was young and strong, I didn't fight. We will continue this fight – as we have no choice, and soon you won't either.'**
> TUVALU YOUTH ACTIVIST

Tuvalu, a beautiful **atoll** in the Pacific Ocean, is home to around 11,000 people. If the temperature and sea levels keep rising, huge numbers of people will have to migrate, but communities don't want to move. Tuvalu communities will have to move to safer places like Australia, away from the home they love.

Tuvalu in the Pacific Ocean

SUCCESS STORIES

Vaan Island, between Sri Lanka and India, has been slowly sinking. This is putting local people at risk as they use the island as a refuge from storms when they're fishing, and it is also damaging the **ecosystem** they rely on. Artificial reefs were added to buffer the waves reaching the island. However, local scientists realised that by replanting the island's seagrass meadows, they could strengthen the island further. Seagrass grows underwater and slows down **erosion**, as well as providing an important habitat for sea horses and fish. These diving scientists are now helping neighbouring islands in the same way.

CHAPTER 3

POLLUTING OUR PLANET

WHAT IS POLLUTION?

Pollution is when harmful materials occur or are introduced into the environment. These materials can be natural or human-made, and can reduce the quality of our air, water and land.

TYPES OF POLLUTION

LIGHT

Light pollution is when there's too much artificial lighting, such as from streets, shops and sports facilities. This brightens the night sky, which affects wildlife's natural behaviours (for example by confusing migrating birds) and scientists' observation of the night sky. Many councils have installed dimming streetlights or shields to focus light downwards to lessen this effect.

NOISE

Noise or sound pollution is when there are unwanted sounds in the environment caused by human activity. These include traffic, roadworks and planes. This can have a health effect on humans such as hearing loss and high blood pressure. It can also affect wildlife – for example, human activities at sea can leave whales and dolphins unable to **echolocate**.

AIR

Air pollution can come from natural causes such as a volcano eruption, but it's mostly human-made. Breathing in smoke, road dust (including tyre particles) and toxic fumes could cause the same damage to a person's health as cigarette smoking.

BREATHING MATTERS
Less than one per cent of the Earth's land area has safe levels of air pollution.

WATER

Many industries, for example mining, use a lot of water, and the toxic chemicals they use can end up in our water. Over a fifth of all known species of freshwater fish have become **extinct** or are **endangered** because of polluted water.

Find out more: waste management pages 38-39, water pollution pages 64-65

What a waste

When we throw something in the bin, it has to go somewhere. If it's not recyclable, it will be burnt or sent to **landfill** sites.

Burning happens in machines known as **incinerators**, and releases toxic gases such as sulphur dioxide and nitrous oxide into the air.

Landfill sites are areas where rubbish is brought from towns and cities. There are half a million landfill sites in Europe. Toxic chemicals and microplastics from the buried rubbish can cause soil and water pollution. Greenhouse gases such as methane are released as the waste breaks down. However, waste doesn't break down quickly…

YOU CAN DO IT!

In nature, there's no such thing as waste. What's not useful to one kind of wildlife, is useful to another. Be a bin-fluencer and see if you can save an item from the bin. What could it be used for instead?

THE WEIGHT OF WASTE
In the US, more than 200 million tonnes of waste are produced each year. That's heavier than 600 Empire State Buildings.

HOW LONG DOES IT TAKE TO BREAK DOWN?

Paper towel
3 weeks

Cotton shirt
5 months

Wool socks
5 years

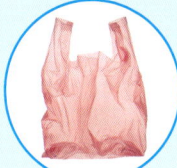

Plastic bag
20 years

Styrofoam cups
50 years

Crisp packets
80 years

Aluminium cans
100 years

Plastic bottles and disposable nappies
500 years

Fishing lines
600 years

Plastic, plastic everywhere

When it was first used in 1869, plastic was a wondrous material: light, cheap to produce, didn't break down quickly and could have many uses. The same qualities make it a problem today.

PLASTIC KILLS
Plastic rubbish in the ocean kills over one million seabirds and roughly 100,000 sea mammals every year.

Find out more: waste management pages 38-39, ocean health pages 56-57

Hard plastic items, such as bowls and lunchboxes, are lighter and safer than glass. If reused and looked after, they can last a long time without ever reaching landfill sites. You may have seen the pictures of a poor sea turtle with a plastic straw in its nose, seabirds mistaking plastic bags for fish or the seahorse clutching a plastic cotton bud. Single-use plastic waste, especially when only used for a few minutes before being thrown away, is a huge problem for the environment.

Plastic is made from fossil fuels, and the extraction of oil and natural gas to make plastics creates a lot of pollution. Plastic eventually breaks up into small particles known as microplastics, which never completely disappear. Being so small means they get everywhere, which makes them difficult to clean up or contain. They have been found in clouds hanging above Japanese mountains, fresh Antarctic snow and human breast milk.

Microplastics are everywhere

PLASTIC DIETS
The average European person who eats seafood also eats around 11,000 plastic particles a year.

 YOU CAN DO IT!

Take part in Plastic Free July and see how many alternatives you can use instead of plastic for a whole month.

All the Rs

We can do so much more with our waste than just throwing it away.

REDUCE

The best course of action is to reduce what we buy. We can choose to buy good quality items that will last longer, and avoid buying single-use or plastic-wrapped items.

YOU CAN DO IT!

In ten seconds, a running kitchen tap can send 1.5 litres of water down the plughole. While 'reduce' usually refers to limiting the things you buy, you can also reduce your water use by…

- Turning off the tap while you're brushing your teeth.
- Having shorter showers or shallower baths.
- Next time you have a wee, seeing if you can 'flush' it away using a jar of water.
- Washing things in a small bowl of water rather than under a running tap.

REUSE AND REPURPOSE

Many items that are thrown away can be reused, for example, freezer bags can be washed out, dried and reused. We can also reuse an object for a different purpose.

YOU CAN DO IT!

Make plastic art using found plastic. This keeps it out of landfill sites and makes people think about plastic waste when they see your creations.

REPAIR

Glue, needles, thread and staplers are some of the simple tools needed to fix broken items.

We can also ask organisations to *replace* plastics in their products with more **eco-friendly** alternatives. We can *refill* containers with cereals, snacks, etc. at certain shops to reduce packaging waste. Finally, turn the page to find out about *recycling*.

Recycling

Recycling is taking something used and remaking it into the same thing or something else. It still uses energy, but far less than creating the items from scratch, and it saves natural resources and produces less pollution.

RECYCLING GLASS

Glass bottles can be recycled over and over again without any loss of quality, but they will take up to a million years to break down at a landfill site.

RECYCLING PLASTIC

Recycling plastic saves oil, energy and carbon **emissions**. However, when different plastics are mixed, it's difficult to separate them before recycling. This often leads to items not being recycled.

RECYCLING PAPER

Recycling paper saves trees, water, oil and electricity. However, paper can't be recycled over and over again. It loses its quality after a while and eventually brand-new tree pulp needs to be added.

If you have paper that has only been used on one side, turn it into a notebook before recycling it. Punch holes in the sides and tie a ribbon or string through the holes to keep it together.

RECYCLING ALUMINIUM

Drink cans, pie cases and baking foil all contain aluminium. Manufacturing new aluminium uses a lot of energy, but recycling it uses 95 per cent less.

There are rules to recycling and you can help those who collect your kerbside recycling by:

- Checking online or asking around about what can be recycled.
- Making sure the correct materials go in the right bins.
- Rinsing out bottles, like the ones used for shampoo and ketchup, before putting them in recycling bins.

CHAPTER 4

THE VARIETY OF LIFE ON EARTH

Biodiversity is the variety of living things found on Earth. It tells us how healthy our planet is. There are around 8.7 million species on the planet, with millions unknown to us in the depths of the vast oceans, lush rainforests and places we just can't get to. The more biodiverse our planet is, the more likely it is that ecosystems will be stable.

Plants are vital. They are the first step in all **food chains** and are known as producers. They give us oxygen to breathe and food to eat, and can store carbon dioxide. Our demanding lifestyles are destroying plant' habitats and leading to species extinction. Once a species is extinct, it is lost forever. This has a knock-on effect on food chains, so others suffer too.

Plants are the first step in a food chain

WHAT IS PHOTOSYNTHESIS?

Chlorophyll is the molecule that gives leaves their green colour and captures sunlight energy. Using this energy, water (taken up by the plant's roots from the soil) and carbon dioxide (taken in through the plant's leaves from the air), the plant can make its own food (called glucose) and oxygen.

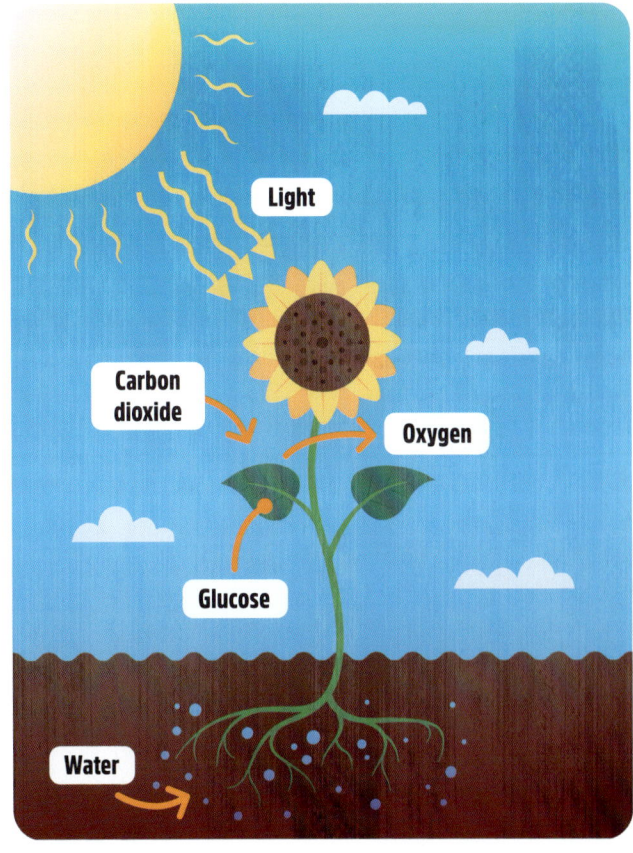

Everyones' home

Habitats are the surroundings that a species needs to survive. Examples include lakes, deserts, **coral reefs**, grasslands and **mangroves**. As more habitats are lost, species have to move or can disappear entirely. Humans living in coastal areas or on islands are losing their homes too.

Mussels, which feed at the bottom of the sea bed, make the water cleaner. The cleaner the water, the more light can get through it and the more sea life thrives. Deep-sea mining is when the seabed is dug up to search for rare minerals used to make the batteries in our phones and electric cars. This destroys the marine habitats of many creatures like mussels, as well as releasing the stored carbon in the seabed. Nature can recover, but it takes time.

Mussels naturally clean water

Rewilding helps to restore ecosystems that have broken down. One way to do this is to introduce **keystone species** that have been lost because of human activities. Keystone species are important in keeping the ecosystem balanced, mainly through food chains.

GOING, GOING, GONE
Three species become extinct every hour.

Beavers are a keystone species

 ## SUCCESS STORIES

In the 1970s, the blue wildebeest in the Serengeti region of Tanzania caught diseases from farm animals and one fifth of the animals died. As the wildebeest weren't there to graze, there was overgrowth of grasses and plants. Overgrowth can easily catch fire and wildfires emit a huge amount of carbon. By reintroducing the wildebeest and managing disease outbreaks, the population recovered. The ecosystem is now back in balance. In less than 10 years, the population of blue wildebeest rose from 300,000 to 1.5 million.

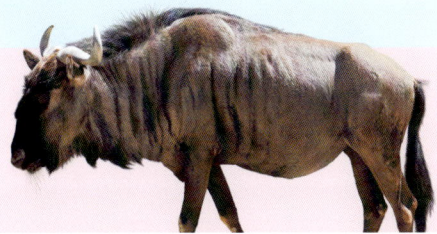

Find out more: deforestation pages 50-51, conservation pages 58-59

Tree-mendous

Trees are wonderful, aren't they? They keep us cool, they are a source of nuts and fruits, and they provide habitats for hundreds of species.

They are life-giving, providing us with oxygen through photosynthesis and absorbing and storing carbon. Yet, every day, tens of millions of trees are chopped down for things like wood, palm oil, chocolate or paper, or to make space for farmland or roads.

This destroys the habitats of animals like orangutans, jaguars, giant armadillos, pygmy hippos and white-bellied pangolins.

TREE-MAIL
In 2015, Melbourne City Council gave email addresses to trees so that the public could report damage. Instead, residents wrote to the trees about how amazing they were.

Orangutan habitats are being destroyed

CATASTROPHIC KICK-OFF

According to the World Wide Fund for Nature, an area the size of five football pitches is cut down every minute in the Amazon.

The most extraordinary forests are the tropical rainforests. They are warm and wet, and home to millions of species, indigenous people and undiscovered plants and creatures. The Amazon rainforest has around 400 billion trees, storing around 150-200 billion tonnes of carbon in the trees and soil. If we carry on with deforestation, these lungs of our planet won't be able to help us limit the rise in global temperatures.

➡ YOU CAN DO IT! ⬅

Next time you pick up a food item with an ingredients label, find out if there's palm oil in it. Can you find a replacement without palm oil or make the food yourself? You can make different choices to help the planet.

Find out more: carbon sinks pages 14-15, biodiversity pages 46-47

Invading our spaces

As humans travel the world, we transport species to new places. This can happen by accident or on purpose, for example, bringing in a plant from abroad to make our garden look pretty or introducing a new fish into a lake because it's delicious. Once these species are in places they shouldn't be, things can get out of control.

Their natural population controllers, such as predators and pests, are not in these new environments; they become **invasive species**, killing off native species and destroying ecosystems. Removing invasive species costs countries millions, if not billions, especially if they affect waterways, roads and farming systems.

Pests can kill plants

A Nile perch

The Nile perch was brought to Lake Victoria, Africa, to boost the fishing industry in the 1950s. While this seemed like a good idea at first, it soon led to the near extinction of the lake's native species such as cichlids, and reduced numbers of shrimps and minnows. Conservation International are working together with other organisations to restore the health of Lake Victoria for native fish to recover.

YOU CAN DO IT!

You can help your local wildlife by:
- Leaving areas outdoors wild, with dead branches and logs for beetles, tall grasses for insects and rocks or stone slabs for slugs and woodlice.
- Feeding birds with seeds and nuts and keeping their feeders clean.
- Planting a mix of flowers (in a garden or in pots on a balcony or patio) to encourage pollinators like bees and butterflies.

Find out more: deforestation pages 50-51, conservation pages 58-59

Losing species

Most of us are in awe of the animals and environment around us, but some humans are greedy.

Poaching and illegal wildlife trade is where species are hunted for their skin, fur, tusks, antlers, organs and much more. These items can be sold for a lot of money. When one species is poached to near extinction, the role it plays within the ecosystem is lost. This can tip the balance and cause trouble for other species, including humans.

Furs are used to make clothing

? WHO'S WHO

Dr Tulshi Laxmi Suwal is co-founder of the Small Mammals Conservation and Research Foundation in Nepal. She works to conserve pangolins, which are threatened by illegal hunting and poaching, as well as habitat loss. Dr Suwal says that successful wildlife conservation is not possible without the participation and support of local communities.

THE IVORY TRADE
It is estimated that 55 African elephants are killed every day for their ivory (which forms their tusks).

There is a valuable trade in ivory carvings

SUCCESS STORIES

Blue whales were wiped out in the Indian Ocean by humans hunting whales for oil (whaling) in the 1960s. In 1978, the Seychelles (the smallest country in Africa) worked to protect the Indian Ocean from whaling. Now, blue whales are returning. Humpback whale populations are also increasing each year. Whales are so important because, during their lifetime, one whale is estimated to capture and store the same amount of carbon as around 30,000 trees. When proper ocean protections are put in place, marine life can bounce back.

A humpback whale

Warmer oceans

As Earth's oceans get warmer, marine life starts to suffer. This includes our precious coral reefs.

Healthy coral reefs are colourful

PLENTY OF SPECIES IN THE SEA
Across 900 islands, the Great Barrier Reef has almost 600 species of coral and over 1,600 species of fish.

The Great Barrier Reef stretches for around 1,430 miles along the coast of Australia. Corals are small animals, similar to tiny sea anemones, and are rooted to the seabed. They live in large groups or colonies, forming a huge structure called a coral reef. Within this are microscopic algae known as zooxanthellae. They give coral their beautiful colours. The algae photosynthesise, providing the coral with food and energy to survive. However, things are changing fast. Since 2014, several heatwaves have hit the Great Barrier Reef, leading to a lot of coral bleaching. Coral bleaching happens when the water gets warmer and algae is pushed out. Without the algae providing their food and energy, the corals bleach or turn white. They are stressed, not dead, but they can die soon after bleaching. Coral reefs across the Caribbean experienced some of the worst bleaching on record in 2023 due to climate change.

Stressed coral turns white

Humans living on the coast need coral reefs for their jobs, food and safety – coral reefs can absorb the force of huge storm waves. Many fish and other ocean dwellers need coral reefs for food, shelter and a place to breed. When only a few corals survive, they can't reproduce and the whole ecosystem collapses. Sea turtles, fish, crabs, jellyfish, starfish and many more organisms depend on coral reefs and may die out too.

Working together

Understanding our place in the world is incredibly important. Humans are a part of biodiversity and we all depend on it.

Conservationists are the people who help to restore balance in ecosystems, protecting animals and their habitats. They have even helped some species come back from the brink of extinction.

 SUCCESS STORIES

The World Wide Fund for Nature works with indigenous people in the Amazon, including the Uru-Eu-Wau-Wau people, helping them record what's happening in the forest using drones. This form of surveillance allows local people to discover newly deforested areas and where fires are occurring. By working together and making use of technology, humans can save rainforests from further damage.

SUCCESS STORIES

In the Himalayan region of Gilgit-Baltistan, livestock farmers used to kill snow leopards to protect their herds. To stop this and revive the snow leopard population, conservation organisations and government agencies have paid farmers since 2006 for any herd losses and built predator-proof pens. When everyone unites, species can survive and not be lost.

SUCCESS STORIES

Back from the brink of extinction, conservationists have bred Australia's white seahorses in captivity. More than 350 were released in 2023 into eight 'Seahorse Hotels' in the Sydney Harbour. The hotels are made of **biodegradable** metal and were positioned a month before. During this month, algae and sponges grew onto the metal, making them much more inviting.

CHAPTER 5
ONE HEALTH

WHAT IS ONE HEALTH?

One Health is an approach recommended by the World Health Organisation. It is all about working together to solve problems, recognising that the health of humans, animals and the environment are closely connected. We may think of ourselves as important, but humans are simply another species on the planet. All species are connected and when the environment is healthy, everyone is healthy.

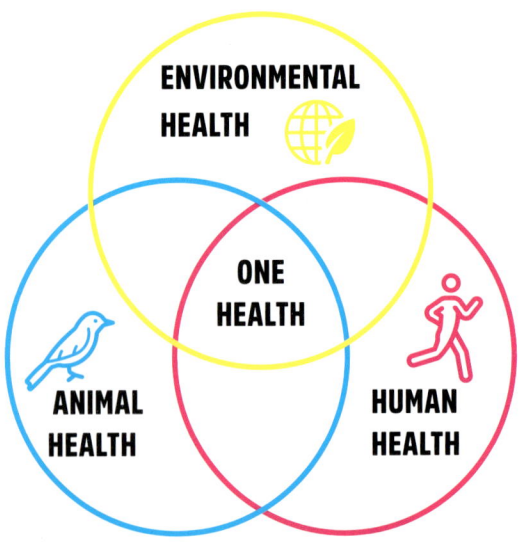

In 2020, we experienced a pandemic where the world shut down. No one on the streets, no planes in the sky, and most shops and businesses closed. Without humans going about their daily lives, we saw a reduction in noise, water and air pollution.

Wearing masks can reduce the spread of disease

Where did *SARS-CoV-2* coronavirus come from? According to some, the virus originated from a laboratory-related accident. However, experts say it's most likely the virus came from an animal, such as the raccoon dog. When human activities (such as logging, mining, farming and building) destroy the homes of wildlife, animals are forced into smaller areas of land. They can become stressed and sick. When a virus jumps the species barrier from animal to human, it is known as a spillover event. Spillover events could become more common if deforestation continues and the planet keeps getting warmer.

One Health is a global public good, and more world leaders are investing in solutions to tackle public health emergencies by understanding our interconnectedness with the environment.

YOU CAN DO IT!

Spread the word about deforestation and volunteer to support a local wildlife charity.

Find out more: pollution pages 36-37, public health pages 62-63

Health products

Keeping yourself clean and well-groomed is good for your health. Things like toothbrushes and soap are important, but we need to understand the impact of them on the environment so that we can make eco-friendly choices. By knowing more about what each product is made of, we know what to do once we're done with it.

There are many eco products available, such as toothpaste powders, bamboo toothbrushes and biodegradable wet wipes, many with **compostable** packaging. However, they might only be compostable or biodegradable on an industrial scale, so we may need to send waste away instead of placing it in a kerbside bin or adding it to our home compost.

SUCCESS STORIES

Blister packs for tablets such as paracetamol are plastic-heavy. Although some chemists offer collection boxes to help recycle these, many end up in landfill sites. Now, there are paper blister packs. These make use of recycled paper with a special, plant-based, compostable coating. They might just be the future for pill packaging.

Not disposing of environmentally-friendly products correctly can be just as damaging to the environment. For example, sending compostable items to landfill can mean that the items are buried in areas with no oxygen so microorganisms can't break them down.

Environmentally-friendly bathroom products

NOT JUST ON CHIPS
Vinegar breaks down dirt, oils, stains and some bacteria. It's safer for the environment than bleach.

 ### SUCCESS STORIES

AKT, Fussy and Wild are small businesses revolutionising the deodorant market. Staying away from aerosols and plastic, they're doing their bit for the environment.

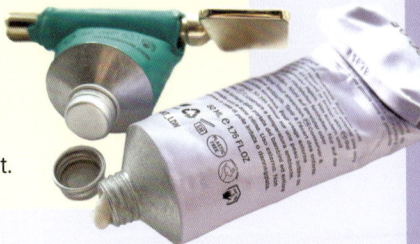

Down the toilet

Weeing and pooing are a part of daily life – everyone does it – but it's where the waste goes that can be a problem for our environment.

In many rich countries, toilet waste is treated to remove anything harmful. Meanwhile, in poorer countries, more than 80 per cent of human waste is sent untreated into rivers, lakes or the sea. The increasing global population leads to more waste, and sometimes it's easier for companies in rich countries to dump sewage straight into rivers too.

In 1957, the River Thames in London, UK was so polluted that it was declared a biological dead zone because there wasn't enough oxygen in the water for wildlife. A successful clean-up campaign led to the return of 125 different species of fish, but the problem isn't solved yet. Since 2020, Thames Water has pumped at least 72 billion litres of sewage into the Thames while making a profit. This has harmed both wildlife and humans, with people having caught serious long-term illnesses from swimming in the river.

Trout and other fish have returned to the Thames

KANGA-POO
The average adult produces and flushes just over 91 kg of poo per year. That's the same weight as an adult male kangaroo.

★ SUCCESS STORIES

Wastewater treatment is an expensive process that uses up a lot of energy. A 2023 Earthshot Prize finalist, Aquacycl, is a company that uses special technology to treat wastewater in an affordable, efficient and less carbon-emitting way. It uses bacteria to make the polluted water safe, while also generating more electricity.

Too hot to handle

With temperatures rising, we may need to get used to heatwaves. Unfortunately, that doesn't mean more paddling at the beach and eating ice cream (as lovely as that would be). Heatwaves can cause a lot of harm to our health, especially for the elderly and really young.

If we don't drink enough water, we can become dehydrated. Too much heat can cause heat exhaustion (dizziness, headaches, confusion and sickness) and this can lead to heatstroke, which can be life-threatening.

It affects wildlife too. African elephants can walk for miles searching for drinking water. As droughts get worse and deserts get bigger, they can't find water fast enough.

MORE TREES, PLEASE
Trees don't just provide shade, they can trap 90 per cent of the sun's heat.

Also, as places get warmer and wetter, disease-causing organisms can migrate and survive. Mosquitoes, bed bugs and various pests are becoming common around the globe, spreading diseases and infesting crops. Sri Lanka has more than 140 species of mosquito, some of which can pass on deadly viruses to humans.

We can learn from countries that have been tackling these pests for years. The larval stage (an early stage of development) of the mosquito needs water. Simple actions such as not allowing water to collect outdoors in flower pots or pet bowls mean that the mosquito cannot lay eggs, so there will be fewer mosquitos.

CHAPTER 6
FOOD AND FARMING

Humans first began farming around 12,000 years ago, when there were fewer people and more space, and the planet had twice the amount of forests we have today. With a continually growing population, there's a greater need for food and farming. This is a problem because trees are cut or burnt down to make space for farmland, so we lose valuable **carbon sinks**, releasing more carbon into the air.

Forests are cut down to make space for farming

FEEDING CLIMATE CHANGE
Food production is the second biggest contributor to climate change, releasing 26 per cent of greenhouse gas emissions.

Farming animals for meat uses up more land, energy and water and releases more greenhouse gases than growing crops. Cows and sheep release a lot of methane (a greenhouse gas) through their burps and farts. Farming methods need to change so that nature can be restored while we continue producing food.

YOU CHOOSE

Producing 1kg of beef releases the same amount of greenhouse gases as farming as 14kg of nuts or 27kg or lentils (which are packed with protein just like meat).

Regenerative farming is a way of growing crops where soil is improved, water is conserved and biodiversity is increased. This in turn helps farming become more productive and profitable.

⭐ SUCCESS STORIES

An Australian company called Sea Forest has created a new feed for cows and sheep, made from a red seaweed known as *Asparagopsis*. Adding just a tiny bit of this to their food can reduce the amount of methane farm animals produce by 90 per cent. Growing more of this seaweed also captures carbon.

Find out more: plant-based diets pages 74-75, sustainable agriculture pages 76-77

Growing to eat

The food we eat often contains ingredients from all over the world. This can give us so much choice and allow us to support poorer farmers abroad through Fairtrade schemes, but getting these foods to us by plane, ship or truck uses up fossil fuels and releases greenhouse gases.

Instead, we can choose to eat local, seasonal and organic food. Local food hasn't had to travel far. Seasonal food is food that is ripe and ready in a particular season – for example, in the UK, strawberries only grow in the summer. Organic food has had no chemicals added to it.

One way to ensure we're eating local, seasonal, organic food is to grow it ourselves. People who don't have a garden can apply for allotments (an area of land) or community gardens to grow fruit, vegetables and herbs.

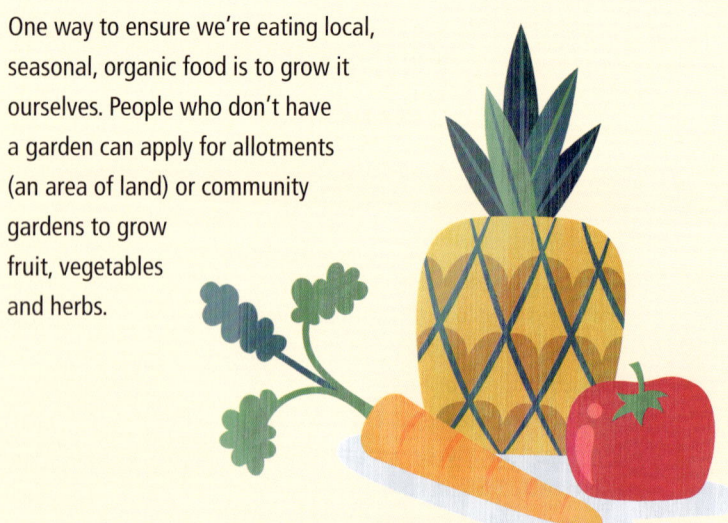

A LOT OF ALLOTMENTS

In the UK, if six people from different households apply for an allotment, their local council has to find them a space to grow food.

Compost is important in organic gardening as it is a natural source of **nutrients**. Almost half of the waste in the average kitchen bin, such as fruit and vegetable peelings, can be turned into compost. Just place it all in a compost bin with some soil and shredded paper and the materials will slowly start to break down. Avoid adding cooked food, or meat and dairy items.

 YOU CAN DO IT!

Try growing tomatoes. You can take seeds out of a shop-bought tomato and plant them into a pot of compost indoors or outdoors.

Love food, hate waste

Sometimes, we put a lot more food on our plate than our stomach can hold, or we only eat the bits of a meal that we like, and we throw away the rest.

Wasting meat is wasting part of an animal, and wasting fruits and vegetables is wasting plants that have been cared for by farmers. In a world where many are going hungry, it doesn't seem right to throw food away.

NEXT STOP, FOOD WASTE
According to the Food and Agriculture Organisation of the United Nations, if food waste were a country, it would be the third-highest greenhouse gas emitter.

There are so many ways we can reduce food waste. We can freeze or refrigerate leftovers. We can find creative, waste-minimising meal ideas online. We can even donate unopened packages and tins to food banks.

Plastic waste in the food industry is also a big issue. Sometimes vegetables need to be covered in plastic to stay protected and fresh. This helps the foods last longer, leading to less food waste. Many supermarkets offer options to buy loose vegetables and fruit, but the plastic-covered versions are usually cheaper. If everyone who can afford it chooses to buy foods that aren't plastic-packaged, maybe supermarkets will have to stop supplying them that way. See pages 40-41 for why single-use plastic is such a problem.

YOU CAN DO IT!

Collect the food packaging that your household uses in one week. Weigh it or spread it all out on the floor and take a photo. See if you buy or use less the following week.

What's for dinner?

Many environmental and nutrition experts suggest eating a balanced diet. They also advise eating less meat and more vegetarian or plant-based dishes throughout the week. This keeps us and the environment healthy because farming animals uses more energy than growing vegetables or grain.

WHO EATS WHAT?

Plant-based individuals only eat foods from plants, such as vegetables, fruits, herbs, flour-based items, lentils, tofu and much more.

Vegans go one step further than plant-based individuals, only eating food from plants and also avoiding animal products entirely (such as not wearing leather or using products tested on animals).

VEGGIE BENEFITS
Eating a vegan diet produces 2.5 times fewer carbon emissions than eating a diet including meat.

Vegetarians eat the same as plant-based individuals, plus honey, eggs and dairy (such as milk).

Pescatarians eat the same as vegetarians, plus fish and seafood.

Omnivores eat the same as pescatarians, plus all meat.

Most vegan and plant-based food companies have the environment in mind so their product packaging tends to be recyclable or compostable too. The ingredients that go into their food (especially at vegan restaurants) are often local and seasonal. However, some plant-based food companies look for flavour and texture above everything else and might not be the best environmentally in terms of food miles and water use.

YOU CAN DO IT!

Why not try the Veganuary challenge which runs every January, or try Meat-Free Mondays and go vegetarian for one day each week?

Something's cooking... or not?

Around three per cent of our home's energy use comes from cooking. Lots of different appliances can cook our meals, but some are more energy-efficient than others.

Gas, electric or induction hob
Most energy-efficient for quick meals as hobs aren't on for too long while cooking

Electric oven
More efficient when used for batch cooking – a lot of food can fit in the oven, so it's great for cooking a lot and storing in the freezer or fridge for reheating

Air fryer
Most energy-efficient when cooking for one to two people or smaller amounts of food

Microwave
Best for reheating meals and quick cooking

Pressure cooker
Best if we need to cook food very quickly

Slow cooker
Most energy-efficient if we are cooking for lots of people and are really busy as these cook food gradually over a long time

There are plenty of other ways we can use less energy when cooking:

- ☑ Fully defrost leftovers in the fridge so they take less time to reheat.
- ☑ Heat water in a kettle before adding it to a pan to avoid using the hob for too long.
- ☑ Use lids on pans when cooking on the hob to trap heat.
- ☑ When baking cakes, put another bake in the hot oven at the same time.
- ☑ Switch appliances off at the plug when they're not in use.

CHAPTER 7

LEISURE AND ENTERTAINMENT

Who do you support? Whether it's football, tennis or athletics, most of us have a team or sports personality we support, but many sports can have a harmful effect on the environment. Popular sports events use a lot of energy, emit tonnes of greenhouse gases and create loads of waste. The sports industry is responsible for emitting around 350 million tonnes of carbon dioxide a year.

SUPER WASTEFUL
The US Super Bowl can generate almost 60 tonnes of waste. That's the same weight as ten adult male African elephants.

Danish footballer Sofie Junge Pedersen who took action

⭐ SUCCESS STORIES

In 2023, 44 players in the FIFA Women's World Cup, led by Danish international Sofie Junge Pedersen, took action for the climate by offsetting carbon (see pages 14–15). The tournament was in Australia and New Zealand, so the carbon emissions from their flights were too much to ignore. They donated money to a tree-planting initiative in Uganda, a habitat restoration project for koalas through WWF Australia and a coastal habitat restoration project in New Zealand.

Some people's idea of sport is actively tipping the food chain balance, like fox hunting. Even though such hunts are illegal in England, Wales and Scotland, they're still happening. Many illegal activities around the world involve animals and are described as a sport, but they're no fun for the animals and the loss of these individuals affects ecosystems.

⭐ SUCCESS STORIES

Forest Green Rovers are the only 100 per cent vegan, **sustainable** and carbon-neutral football club in the world.

Going online

We spend a lot of time online. Gaming, streaming films, catching up on TV shows, video-calling and sending emails and messages all need Wi-Fi or a mobile network, and all use energy.

100 PER CENT
Most mobile phones become fully charged within two hours. In the UK, almost £50 million is wasted every year by charging devices overnight.

YOU CAN DO IT!

- Switch off and unplug appliances when they're not in use – appliances can still use electricity when plugged in. Standby lights can still be working and this is wasted energy.

- Buy a refurbished phone instead of a new one. Refurbished phones reuse electronics rather needing new metals and natural resources. Extracting those is a process which emits carbon. There's also less waste as you're reusing a phone instead of throwing it away.

- If your laptop is running slowly, ask a technician to upgrade it. Adding memory or updating software will cost and waste less than buying a new laptop.

Data centres keep our online information, including our emails, safe. They are made up of storage security systems and servers. They use a lot of power and produce a lot of heat. In the US, they're cooled down by around 300,000 gallons of water per day which is stored in cooling towers. This wastes a lot of water, but these setups can also lead to disease outbreaks. Europe has used 'dry' cooling towers for decades, which conduct heat through air-cooled heat exchangers. As there is no direct contact between the water and the air, there is no water loss or disease risk.

A data centre cooling tower

DELETED EMAILS

If everyone deleted ten emails, it would save around 55 million kilowatts of power – enough to power 5,000 homes for a year.

Find out more: energy use pages 20-21, electronic waste pages 38-39

Time to play

Toys, board games and puzzles are great for family time and developing our imagination and skills, but how do they impact the environment?

The toy industry uses the most plastic of all industries. 90 per cent of toys are made of plastic because it's light and easily mouldable, and can be made to be squishy or brightly coloured. When these toys are thrown away, they can end up in landfill sites and cause harm to the environment.

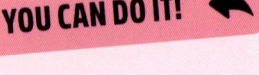 YOU CAN DO IT!

Instead of buying, try borrowing. Ask neighbours and friends at school, or find out whether there is a Library of Things or a Toy Library near here you live. This is much kinder to the planet if you decide you don't need something after trying it out.

Even when jigsaw puzzles and wooden blocks appear to be made from natural materials, they might not be made from sustainable wood (wood from a place where trees can be cut down and replanted with little effect on the local ecosystem). Look out for the FSC symbol on packaging; it shows that the materials come from responsibly-managed Forest Stewardship Council-certified forests. The extra information under the logo tells us whether the material is recycled or a mix.

Some toys and games need batteries and throwing these away in household waste can cause harmful acid to leak into the environment. Some local grocery stores and supermarkets offer places for us to leave our unwanted batteries for them to dispose of safely.

RECHARGING
Regular batteries become waste after one use, but rechargeable batteries can be used and recharged up to 1,000 times.

Celebration time

Whether it's Eid, Diwali, Hanukkah, Christmas, birthdays or weddings, we all enjoy celebrating with our loved ones, but the material side of these celebrations isn't always great for the environment.

Wrapping paper that is made with glitter or foil coatings can't be recycled. Balloons released into the sky are often choked on by wildlife when they land. Fireworks contain chemicals which can dissolve in water, contaminating our rivers and lakes; they also affect our air quality by releasing smoke and other harmful particles. Disposable plastic crockery is easy to transport and can be thrown away without the need for washing up, but over 90 per cent of this plastic ends up as litter or in landfill sites rather than being recycled.

 SUCCESS STORIES

In Rajasthan, India, Bartan Bank (Utensil Bank) has been set up so villagers can borrow steel crockery for any event, for free.

THAT'S A WRAP
227,000 miles of wrapping paper is used each year in the UK. That's enough to wrap around the Earth nine times.

 YOU CAN DO IT!

Celebrate in style by giving loved ones eco-friendly gifts:
- Choose recycled cards and wrapping paper or reuse newspaper to wrap gifts.
- Instead of wrapping paper, use a reusable gift bag.
- Make edible gifts and decorations, like a jar of homemade cookies.
- Upcycle by making gift tags out of old, cut-up cards.
- Buy gifts from ethical companies that don't use plastic packaging.

CHAPTER 8
FAST FASHION

Going shopping for new clothes and accessories can be fun, but the fashion industry causes a lot of damage to our environment.

Creating clothes uses a lot of energy and water; many of the chemical dyes used are toxic which can pollute our freshwater. Many fabrics are also a source of microplastics as they break down in landfill sites, or release toxic fumes when burnt.

TOO FAST
In 2018, the Clean Clothes Campaign found that three out of five fast fashion items end up in landfill sites within a year.

WHAT IS FAST FASHION?

Fast fashion is when clothes are produced on a massive scale, quickly and cheaply, so customers feel they can buy whatever and whenever they want to keep up with the latest trends. Unfortunately, it has a huge impact on the environment.

 YOU CAN DO IT!

It's easy to avoid fast fashion. Try looking for second-hand and vintage clothes in charity shops to show off your own style of eco-fashion. Win-win: you get cool clothes and a charity gets money.

Find out more: waste reduction pages 38-39, sustainable materials pages 88-89

Natural vs synthetic

Natural materials are anything that comes from plants, animals or the earth. Synthetic materials are created artificially by humans.

Natural materials usually have less of an environmental impact as they use fewer chemicals during the production process and are more likely to be biodegradable and recyclable. However, some natural materials such as linen, cotton and bamboo also need lots of water to grow.

Cotton plants

THIRSTY WORK
It takes 2,700 litres of water to grow the cotton needed for one T-shirt. That's the same volume as 10,000 mugs of hot chocolate.

It's important to make sure that the materials you're wearing are sustainable. Sometimes, synthetic materials are more sustainable than natural ones.

SUCCESS STORIES

- Dr Raquel Prado and her team at Ananas Anam have created Piñatex, turning fibres from waste pineapple leaves into clothing and shoes.
- Finland's Spinnova has invented a fibre that is just like cotton but uses 99 per cent less water to produce. There are no toxic chemicals and it can be recycled over and over without losing its quality.
- Synthetic fabric dyeing uses water equal to two million Olympic-sized swimming pools every year. Orr Yarkoni and Jim Ajioka set up Colorifix, which uses the genetic codes of natural plant dyes. This colouring system saves water, energy and chemical pollution.

Fabric dyed with natural dyes

YOU CAN DO IT!

Choose clothes made from sustainable materials that you will wear for a long time on many occasions. Also, try on clothes in-store instead of ordering several items online and having to return them – more deliveries mean more carbon emissions.

Glitz and glamour

Sparkly outfits often make appearances at parties and concerts, fancy events, Christmas and New Year. While this glitz and glamour may look incredible, glitter, shiny beads and sequins are all tiny pieces of plastic with metallic coatings which rely on fossil fuels to be made.

These sewn or glued-on items often fall off as we move or in the wash. Once they're released into the environment, they'll be there forever, breaking down into tinier pieces and never quite disappearing.

While biodegradable sequins have been invented, they are not yet mass-produced, likely because they're more expensive to make and less robust. Plastic sequins are punched out of sheets of plastic, the remains of which create even more waste. Some of this is burnt, releasing toxic smoke and fumes into the air.

DUMPING GROUND
40 per cent of items made by the clothing industry are never sold.

The European Union (EU) has banned the use of glitter, but it's still available in some countries outside of the EU like the UK. Perhaps, as customers, we should just stop buying it.

'TIS THE SEASON
95 per cent of all Christmas jumpers are made of acrylic, a synthetic material which sheds microplastics every time it's washed.

 YOU CAN DO IT!

The charity Oxfam found that only 25 per cent of people would wear their Christmas jumpers and party outfits again. Some companies now offer rental services; try to rent party outfits instead of buying them.

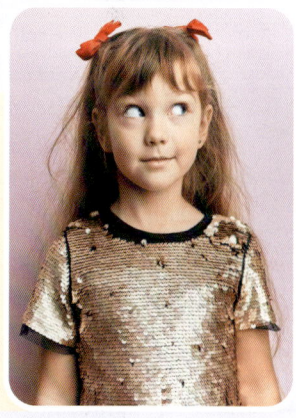

Find out more: sustainable materials pages 88-89, recycling pages 44-45

Don't do the laundry

Changing the way we wash and dry our clothes can help us use less energy and water. Unless clothes are really stinky and stained, they don't necessarily need to be washed. Over-washing makes clothes shrink, fade and fall apart, and when this happens, they often get thrown in the bin and sent off to landfill sites.

Levi's, a clothing brand known for making denim jeans, says we only need to wash jeans after every 10 wears. Wool naturally keeps dirt and smells at bay, so woollen items only need to be washed every 5 wears or so.

★ SUCCESS STORIES

Soap nuts are the dried shells of soapberries. They are grown in India and contain saponins. These are soap-like chemicals and can be used to make natural detergents. Three soap nuts placed in a small cotton bag can be added to laundry instead of washing powder or other detergents and reused five to six times.

MAKE EACH WASH COUNT

A large washing machine uses 160 litres of water for a single load. That's double the amount of water in the average bath.

YOU CAN DO IT!

There are plenty of ways you can make your washing more environmentally friendly:
- If you desperately need one item of clothing washed before the family laundry is done, hand wash it in a small bowl of lukewarm water with a teaspoon of eco-detergent.
- Save energy by washing at lower temperatures (30°C or 20°C) or using the eco-setting on your washing machine
- Save even more energy by hanging washing outside to dry if you can rather than using a tumble dryer.

A second life

When clothes are ripped, stained or need alterations, it can feel easiest to just throw them away and buy something new, but clothing repairs are usually quick and easy, and save money and the environment.

Many people throw clothes away because they don't know how to mend them. If you don't know how to add a missing button, hem or patch, or to fix a broken popper, try asking around in your local community or watching videos online. There may even be a repair café or sewing shop near you.

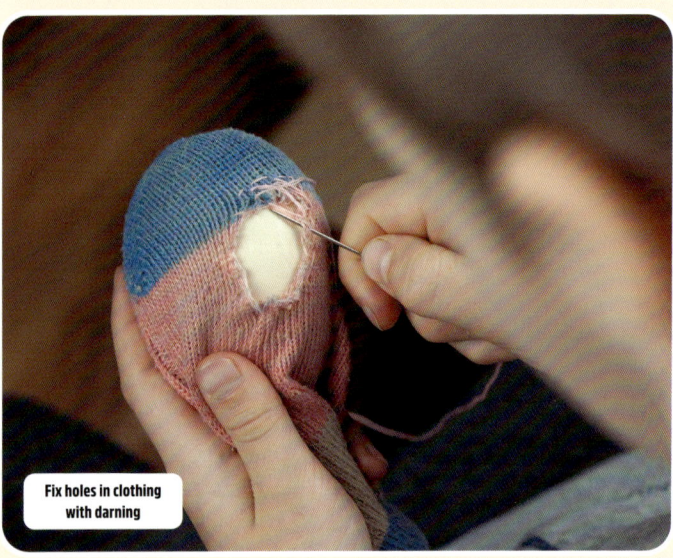

Fix holes in clothing with darning

SUCCESS STORIES

Around 7,000 tonnes of clothing are thrown away in France each year, with much of it ending up in landfill sites. In the summer of 2023, the French government announced a plan for citizens to get their clothes repaired at low cost or for free. *Refashion* is the eco-organisation behind it, and the aim is to create a circular economy, ensuring that items that would have been thrown away get reused or fixed so there's no waste.

Sashiko embroidery

YOU CAN DO IT!

Mending doesn't need to be dull or boring, it can be quite decorative and trendy. *Sashiko* is a form of functional embroidery from Japan. Historically, it was used to strengthen and cover up areas of wear and tear on clothes. Look it up and give it a try, or add an embroidered patch to your clothes instead.

CHAPTER 9
TRAVELLING AROUND

One-fifth of greenhouse gases come from transport. There are 1.4 billion vehicles (cars, vans, lorries etc.) in the world, most of which are powered by petrol and diesel. Petrol and diesel are made from fossil fuels. As vehicles burn these fuels to create power to move, harmful waste gases leave through the exhaust system, polluting the air we breathe. Vehicles are fitted with catalytic converters to reduce the amount of these gases, but carbon dioxide and other pollutants can still make their way into the atmosphere.

Short car journeys are energy inefficient because the engine doesn't have time to warm up and so uses up more fuel. If we do need to use a car, 'carpooling' or sharing lifts so that all seats are occupied leads to less carbon being emitted per person and means there are fewer cars on the road.

It's not just land travel that causes problems. Every day, there are over 100,000 flights carrying passengers to far-off places. Other flights carry goods around the globe. Planes emit more greenhouse gases per passenger per kilometre than cars, buses or trains. They release these gases while up in the air, which has a more harmful greenhouse effect than when gases are released at sea level. Despite this, more and more people are choosing to travel by plane because of the cheap flights offered by airlines looking to make more money.

EMISSIONS PER PASSENGER PER KILOMETRE TRAVELLED

Transport	CO_2 emissions	Secondary effects
Domestic flight ✈	133g	+121g
Long Haul flight ✈	102g	+93g
Car (1 passenger) 🚗	171g	
Bus 🚌	104g	
Car (4 passengers) 🚗	43g	
Train 🚆	41g	

- CO_2 emissions
- Secondary effects from high altitude, non-CO_2 emissions

ALL ABOARD
Travelling by train releases around 75 per cent fewer emissions than travelling in a petrol- or diesel-fuelled car.

Find out more: fossil fuels pages 16–17, air pollution pages 36–37

Off on holiday

Taking a break from the hectic timetable of school, work, clubs and busy weekends is always welcome, especially when we can travel somewhere new.

Going on holiday close to home or by coach, train or ferry can be much better for the environment than flying which causes a high level of greenhouse gas emissions.

Going abroad isn't all bad – countries in the **Global South** need small-scale, sustainable tourism to bring in money. However, they need travellers who want to learn, to live like the locals and respect the land.

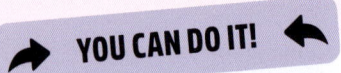

YOU CAN DO IT!

If you're on a beach holiday:
- Ask a grown-up to help you do a beach clean if you see any litter or plastic pieces.
- Remember that shells and pebbles aren't free souvenirs – leave them on the beach as useful materials for wildlife habitats and for good environmental health.

WHAT IS ECO-TOURISM?

Eco-tourism is nature-based tourism where visitors travel responsibly to natural conservation areas, which helps the local communities and raises awareness of environmental issues.

 SUCCESS STORIES

In 2001, Roque Sevilla, the former mayor of Quito, Ecuador, bought part of the Chocó Cloud Forest known as the Mashpi reserve. It's a place for tourists to stay while learning about the vast biodiversity of the forest. The Mashpi reserve has helped local people move away from jobs in deforestation by providing them with alternative careers, such as becoming experts in conservation.

Slow travel

Instead of flying from London to Bologna, or from Melbourne to Perth, what if we took a train or coach? The green option is not always the cheapest or fastest, but we may enjoy taking the time to admire the natural scenery and know that we're creating fewer emissions.

 SUCCESS STORIES

To attend climate conferences in New York, Greta Thunberg famously travelled across the Atlantic on board a carbon-neutral racing yacht called Malizia II. There wasn't a proper bed, toilet, shower or kitchen on board, and the trip took 14 days.

⭐ SUCCESS STORIES

Ed Gillespie, futurist and co-founder of Futerra, travelled around the world without flying from 2007–2008. He didn't use traditional methods of non-flying transport either, travelling by kayak, cargo ship, hovercraft and camel.

On a smaller scale, cycling is also a great way to slow travel. It's faster than walking, and it's a green and healthy way to get around. Cities around the world are gradually changing to become more bike friendly.

❓ WHO'S WHO

Tahsin Uddin

a climate activist in Bangladesh, encourages people to cycle. He set up a youth organisation called Lal Sabuj Society to encourage young people to cycle, plant trees and organise clean-ups. Young girls in Bangladesh are often not taught or allowed to ride a bicycle, but Tahsin's organisation has trained over 100 girls to cycle.

Find out more: transport emissions pages 96-97, sustainable tourism pages 98-99

To infinity and beyond

Humans love knowledge. We explore the planet by climbing the highest mountains, wandering in the darkest rainforests and diving into the deepest oceans. We still don't know everything about our Earth, but many are already looking beyond our planet.

One per cent of the world's population is super-rich. These people create thousands of times more carbon emissions than the average person through their yachts, private jets and what they invest in. This has been made worse by the invention of space tourism.

Blue Origin NS-18 was a sub-orbital spaceflight mission. In 2021, this mission made 90-year-old Star Trek actor, William Shatner, the oldest ever space tourist. It cost $5.5 billion for him to travel 60 miles above the Earth's surface and experience four minutes of weightlessness.

Virgin Galactic has done a similar flight in a VSS Unity rocket plane. Per passenger per mile, the Virgin Galactic trip emits 60 times the carbon of a commercial transatlantic flight. Black carbon, or soot, is also released during rocket launches.

SPACE RACE
In 2022, the global spending on space exploration was around $103 billion.

Would you go on holiday to space?

Perhaps because space tourism is still fairly rare, we needn't worry too much about it, but if it does become common, the environmental impact could be very damaging. Would you be a space tourist?

SPACE JUNK
There are over 130 million pieces of space rubbish orbiting Earth, including abandoned spacecrafts and broken parts of rockets.

Fuelling our travel

Petrol and diesel are both made from fossil fuels. See pages 16-17 for why using fossil fuels is bad for the environment.

Looking for alternative fuels is really important. Biodiesel is a renewable fuel made of anything from used cooking oil and animal fats to plant matter, algae and sewage. Using biodiesel reduces emissions by two-thirds compared to petrol-based fuels.

? WHO'S WHO

Dr Florence Gschwend
a Swiss chemical scientist, discovered a low-cost way to turn waste wood into renewable biofuels.

POO BUS
In 2015, a bus in Bristol, UK, ran on biomethane gas from the human and household waste of 32,000 homes.

Buses go green

 ## SUCCESS STORIES

- In 2023, a Virgin Atlantic commercial plane running on plant sugars and waste fat flew from London to New York. No fossil fuels were used. This Sustainable Aviation Fuel emits 70 per cent less carbon than petrol-based jet fuel. This is a great start, but it's expensive to run all flights this way. The airline industry needs to develop fuel and battery technologies quickly to become completely zero emission.

- In 2023, over a third of fuel used at European motorsport races was made from Hydrotreated Vegetable Oil (HVO). This is a fossil-free diesel made from waste oils and plant matter, and can reduce carbon emissions by up to 90 per cent. HVO fuels are being increasingly used for city council waste collection trucks and delivery vehicles.

A rally car at a European race in 2023

What can we change?

Although you probably don't have a licence to fly or drive, you can still make changes to your travel choices, and encourage others to do the same. If possible, we can choose not to use a car. Some cities combat smog by having car-free days.

 YOU CAN DO IT!

Encourage your friends, family or school to take part in World Car Free Day on 22nd September every year.

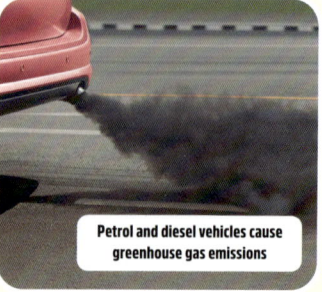

Petrol and diesel vehicles cause greenhouse gas emissions

If a car is needed, we can choose to drive one that is better for the environment. In 2017, the French government announced that the sale of petrol and diesel vehicles will be illegal from 2040, and other countries are making similar laws.

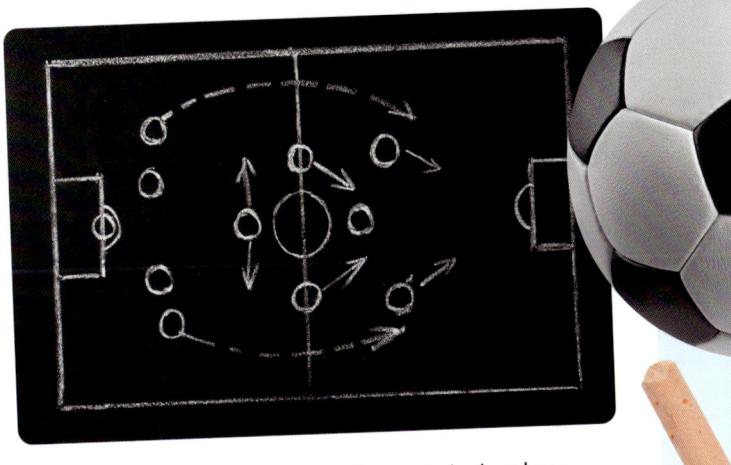

Imagine playing a football game. If we only look at how many goals our team scores, we might think we're winning. We also need to look at how many goals we're letting into our own net, because we might actually be losing.

The Dasgupta Review published in 2021 calls for changes in how we think, act and measure success as countries. If we only measure economic growth to see if we're successful then we'll continue to damage the environment. so we need to think about success differently.

TIPPING THE SCALES
Human-made materials now outweigh Earth's entire biomass.

We can look at economic growth as one sign of success, but we also need to consider the loss of nature that was needed to achieve that growth. Maybe the economic growth wasn't such a success after all. Instead, we could try measuring success by how healthy our planet is.

Find out more: greenwashing pages 114-115, sustainable business pages 110-111

Banking on money

Where does our money actually go when we put it into a bank? A small amount is held as cash at the bank, but the rest goes towards loans. This is when banks lend other people and businesses money.

SUCCESS STORIES

Some banks use our money for good. Starling Bank in the UK is branchless, paperless and runs on 100 per cent renewable energy. They support companies working on green transport and sponsor climate change projects. When a customer gets a friend to join, their partner, Trillion Trees, will plant a new tree. Other banks such as the Netherlands's bunq, Germany's Tomorrow and Italy's Flowe support similar environmentally-friendly projects.

Around the world many people of all ages and backgrounds are joining together with environmental groups to protest peacefully their frustration at banks and companies that support the fossil fuel industries.

PROTEST

In 2019, 50 of the largest banks lent $2.6 trillion to companies destroying nature. Barclays and HSBC were the top European funders of fossil fuel industries between 2016 and 2022.

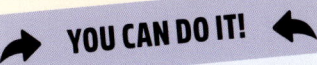

YOU CAN DO IT!

Before putting money in a bank, make sure you've looked into what they're going to do with your money. If everyone chose an environmentally-friendly bank, there would be a lot less money going towards harming the planet.

Find out more: corporate responsibility pages 114-115

Our homes

The human population has tripled over the last 70 years, and every global citizen needs somewhere to call home. Building more houses leads to more land being used. Trees are often lost in redevelopment, so there are fewer carbon sinks, especially if new trees aren't planted.

Lots of common building materials are damaging to the environment. Steel uses a lot of fossil fuels during production and is responsible for eight per cent of carbon emissions caused by humans worldwide. Cement is a binding material that causes a huge rise in carbon emissions too. If we want to save our planet, we need people in power to invest in and promote more sustainable building practices.

Steel, concrete and cement can all damage the environment

Sustainable societies, such as in Sweden, build houses out of wood. Building with wood is good for the planet if the timber is from sustainably managed forests. Wood can be as strong as concrete or steel but lighter to transport. Wooden buildings act as carbon sinks too, storing more carbon when they're built than they release into the air.

ECO HOUSES
Ecocapsule Space is a tiny egg-shaped living space that runs on solar and wind power.

The problems don't stop when a house is built; environmentally-friendly choices need to be made by the owner:

- ☑ Keep gardens and outside space wildlife friendly by not removing grasses or plants, not adding concrete, and not adding fake lawns.
- ☑ Buy appliances that have the highest energy efficiency rating (A+++).
- ☑ Choose triple glazing for windows to keep more warm air in and cold air out.

Find out more: energy efficiency pages 20-21, sustainable materials pages 88-89

Greenwashing

No company wants to look evil, do they? Companies involved in planet-damaging activities might shine a brighter light on some small, eco-friendly activities they're doing instead. This is known as greenwashing. They're showing themselves off as a solution to the climate crisis, while actually doing harm elsewhere.

When a company whose main business is extracting oil and gas also runs a competition for students to design and race energy-efficient cars, it doesn't seem to make sense. This is an example of greenwashing because their main business is damaging to the planet but they're trying to associate themselves with environmental concerns.

SUCCESS STORIES

German forester, Peter Wohlleben, looks after forests in an eco-friendly way. He doesn't use insecticides or heavy machinery, and the trees are cut by hand and moved by horses. Taking a more natural approach to forest management means a healthier forest, which lives on for decades. Healthy trees also means that when timber is needed, it's good quality. The forest was losing money before he changed things around.

Greenwashing is becoming so widespread that we're unable to see the ethical companies actually working to save the planet – but they're out there. Many companies are stuck in their ways, but old systems aren't necessarily the best for the planet or for profits. All companies can find a way to make profits and help save the planet.

  ## YOU CAN DO IT!

If you don't like something, you don't have to support it. If you know of a festival, concert or event being sponsored by a company that encourages environmental damage, you can choose not to attend. If the organisers know that the public isn't happy, they will have to change.

Support events with values you believe in

Find out more: sustainable business pages 110-111, consumer power pages 114-115

Fighting for the planet

Organisations like Greenpeace, WWF and Client Earth are fighting on our behalf so that we'll always have a liveable planet. Companies like Wild Justice take on legal cases on behalf of wildlife, working hard to hold companies accountable for deforestation and the loss of biodiversity.

Change needs to happen now if we want to protect our future. Most people think they can't do much to help the planet, and young people might feel especially unable to make a difference. It's important to remember that we can all speak up. We can all help, no matter where we live or how old we are. The two biggest causes of climate change are burning fossil fuels and **agriculture.** We must demand that governments and businesses give us clean energy and products, instead of chasing after profits.

❓ WHO'S WHO

Txai Suruí
founder of the Movement of Indigenous Youth of Rondônia, is an activist trying to save her home. Rondônia, in Brazil, has been affected by the climate crisis due to extreme deforestation. Rivers are running dry and species are disappearing fast. Txai Suruí studied law and now works with the Kanindé legal team to protect the rights and land of indigenous people.

➡ YOU CAN DO IT! ⬅

Write to your local politician or ask a grown-up to help you sign petitions. You can make your voice heard by those who have the power to change things. You deserve better from those who are meant to keep you safe.

Find out more: activism pages 4-5

FIND OUT MORE

We live on a wonderful planet, full of amazingly beautiful places, people and wildlife, so let's take care of it. If you'd like to find out more about climate change and the environment, here are some useful websites:

Big Butterfly Count: www.bigbutterflycount.butterfly-conservation.org
BioBlitzes: www.nhm.ac.uk/take-part/citizen-science/bioblitz.html
Citizen science projects with Zooniverse: www.zooniverse.org
Client Earth: www.clientearth.org
Conference of the Parties (COP): www.unfccc.int/process/bodies/supreme-bodies/conference-of-the-parties-cop
The Earthshot Prize: www.earthshotprize.org
Freecycle: www.freecycle.org
Friends of the Earth: www.friendsoftheearth.uk
Fridays For Future: www.fridaysforfuture.org
Greenpeace: www.greenpeace.org/global
Love Food, Hate Waste: www.lovefoodhatewaste.com
Recycle Now: www.recyclenow.com
Royal Society for the Protection of Birds: www.rspb.org.uk
The Story of Plastic: www.storyofstuff.org/storyofplastic
The World Wide Fund for Nature (WWF): www.wwf.org.uk

GLOSSARY

Activist: a person who campaigns or protests to bring about change.

Agriculture: the science and practice of farming.

Algae: a group of aquatic organisms that can photosynthesise.

Asteroids: large rocks that orbit the Sun (if a piece of one lands on Earth, it is called a meteorite).

Atmosphere: all the gases that surround the Earth.

Atoll: a ring-shaped island or reef (or set of islands) made of coral.

Biodegradable: can be broken down naturally by bacteria or fungi, but sometimes very slowly and leaving behind materials that are not good for the soil.

Biodiversity: the variety of living organisms found on Earth.

Carbon dioxide: a major greenhouse gas that living organisms breathe out and is released when fossil fuels are burned.

Carbon sink: anything that absorbs more carbon from the atmosphere than it releases, like plants, the ocean and soil.

Climate change: a long-term change in the climate (average weather conditions over a long time).

Compostable: something that will break down over time into nutrients, like vegetable peelings.

Conservation: the protection of a natural resource, habitat or species.

Coral reef: a large structure in the sea made from a build-up of sea creatures known as coral.

Deforestation: the cutting down of many trees for land, fuel or wood.

Drought: a long period of very dry weather.

Earth's crust: the outer layer of planet Earth, on which there is land and oceans.

Echolocate: the way some animals find objects or prey by using sound waves that are reflected back to them.

Eco-friendly: describing something environmentally-friendly, or not harmful to the environment.

Ecosystem: an area containing communities of living organisms, their homes and their surroundings.

Emissions: released gases or chemicals.

Endangered: a species that is close to extinction.

Erosion: when soil is worn away and transported somewhere else by wind or water.

Evaporate: when a liquid turns into a gas.

Evolve: when living things develop or gradually change over time by passing beneficial features from parent to child.

Extinct: when a species dies out completely, or the few still alive can no longer reproduce.

Fault lines: long cracks in the surface of the Earth.

Fertiliser: a natural or artificial mix of nutrients to add to the soil to make plants grow well.

Food chain: a diagram that shows how energy passes from one living thing to another in an ecosystem, based on who eats what.

Fossil fuels: a natural fuel formed underground from animals and plants that died millions of years ago.

Fracking: the process of pumping water, sand and chemicals deep underground at high pressure to cause cracks in the rock, releasing the oil or gas trapped inside.

Generator: a machine powered to produce electricity.

Global North and Global South: the Global North describes developed countries and the Global South describes developing countries. Developed countries tend to have more economic and political power across the globe, while developing countries have less – there are many (often unjust) historical reasons for this.

Global warming: a gradual warming of the Earth's average surface temperature caused by human activities.

Green: used to describe eco-friendly choices, policies and behaviours.

Greenhouse effect: the natural trapping of heat in Earth's atmosphere.

Greenhouse gas: gases in the atmosphere (like carbon dioxide and methane) that cause the greenhouse effect.

Habitat: the natural home or surroundings that a living thing needs to survive.

Ice age: a long period in Earth's history when very cold temperatures resulted in ice forming.

Incinerators: machines used to burn waste.

Indigenous: people who are the descendants of the earliest inhabitants of an area.

Industrial Revolution: a period in the 18th and 19th centuries when machines were more widely used in Europe and the US to make work easier for people.

Insulated: preventing the loss of heat.

Invasive species: living organisms that have come to a new area and cause damage to the surrounding environment.

Keystone species: if this species is missing, the rest of the ecosystem will collapse.

Landfill site: a huge area of land where rubbish is dumped straight onto the ground or into a large hole. The waste is spread and flattened, with other materials (like soil) added on top, ready for the next layer of rubbish.

Mangrove: groups of trees and shrubs that grow in coastal waters.

Microplastics: extremely tiny pieces of plastic.

Mining: the extraction of natural materials from the surface of the Earth.

Net zero: when there's a balance between the amount of greenhouse gases emitted and the greenhouse gases removed from the atmosphere.

Nutrients: the important chemicals, like carbohydrates, proteins, vitamins and minerals, found in food that help living organisms grow.

Organism: an individual animal, plant or single-celled life form.

Oxygen: a gas found in the air that animals breathe in to live.

Phase out: to stop completely.

Poaching: the illegal killing of wild animals to make money.

Policy: a set of ideas or plans used to make decisions, most often used by governments and businesses.

Radioactive: emitting ionising radiation (a type of energy).

Regenerative farming: a way of growing crops involving fewer artificial chemicals, not disturbing the soil too much, and growing different types of crops in the same area across different seasons.

Renewable: from a source that does not run out, like wind and sunshine.

Species: a group of living organisms that are scientifically similar to one another.

Standby: a setting where a device looks like it's off, but is actually on and still using energy.

Sustainable: continuing over a very long time with little or no harm to the environment.

Tonne: a measurement of weight equal to 1,000kg.

Turbine: a device that spins using the movement of wind or water to generate electricity.

United Nations: the global organisation encouraging countries to work together to solve global issues.

Water cycle: the continuous movement of water through the Earth and its atmosphere.

Wildfire: when a fire burns out of control across a huge countryside area.

Electrical cars can be charged using renewable electricity

Electric vehicles are a great alternative. There were over 14 million electric cars on the world's roads in 2023. Running an electric car produces no harmful air pollution and, unlike a petrol or diesel car, they can be completely greenhouse-gas free if the charging-point electricity comes from renewable sources. However, if electric vehicles are the way forward, then battery life and charging times need to be improved, and more charging points need to be installed.

 ## SUCCESS STORIES

The Mayor of London, Sadiq Khan, introduced the Ultra Low Emission Zone (ULEZ) in 2019. ULEZ helps clean London's air by taxing (and so reducing) older, more polluting vehicles on the road. ULEZ has been so effective at improving the air quality that it was expanded to include all London boroughs in 2023.

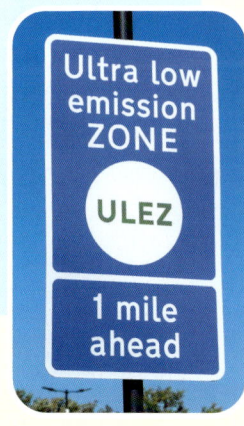

CHAPTER 10
MONEY AND POWER

Many years ago, humans worked on the land and were self-sufficient, living completely on what they grew and made themselves. In some places around the world, life is still like this, but most of us depend on others to grow our food, make our clothes and provide us with water, light and heating. We earn money from our jobs to pay for these. Many people, businesses and governments take nature for granted and focus instead on making money.

Governments want to increase the wealth of their own countries, but they often do this at the expense of the environment. Can we save the planet and be wealthy countries? Yes, we can.

IMAGE CREDITS

Front cover and pi (tr) HuHu/Shutterstock; (mr) Macrovector/Shutterstock; (br) TWINS DESIGN STUDIO/Shutterstock; vi Dr Tushi Laxmi Suwal, used with permission; p1 David Pereiras/Shutterstock; p2 chaiyapruek youprasert/Shutterstock; p3 studiovin/Shutterstock; p5 Ben Gingell/Shutterstock; p6 Jakub Rutkiewicz/Shutterstock; p7 SewCreamStudio/Shutterstock; p9 (tl) Alaskagirl8821/Shutterstock; (tr) Alexandros Michailidis/Shutterstock; (ml) Wirestock Creators/Shutterstock; (mr) Heidi Besen/Shutterstock; (b) sanddebeautheil/Shutterstock; p11 Sharomka/Shutterstock; p12-13 David Koscheck/Shutterstock; p14-15 (bg) krstrbrt/Shutterstock; p15 A3pfamily/Shutterstock; p17 (t) xpixel/Shutterstock; (b) Aleksandr Gogolin/Shutterstock; p18 Sherif Ashraf 22/Shutterstock; p19 (t) Rawpixel.com/Shutterstock; (b) Tomas Ragina/Shutterstock; p20 MNStudio/Shutterstock; p21 (lightbulb icon) The Studio/Shutterstock; (other icons) Laura Neate, used with permission; p22 fuyu liu/Shutterstock; p23 (t) fokke baarssen/Shutterstock; (b) nikonka1/Shutterstock; p24 Scharfsinn/Shutterstock; p25 (t) Breedfoto/Shutterstock; (m) Kletr/Shutterstock; (b) TonyV3112/Shutterstock; p26-27 (weather icons) Laura Neate, used with permission; p26 (b) Viacheslav Lopatin/Shutterstock; p28 (m) Ruben Perez Gil/Shutterstock; p30 Doga Ayberk Demir/Shutterstock; p31 (t) Joe Barti/Shutterstock; (m) Paul Harrison/Shutterstock; (b) Marinel Ubaldo, used with permission; p32 Lua Carlos Martins/Shutterstock; p33 (t) MDay Photography/Shutterstock; (b) Jakub Maculewicz/Shutterstock; p35 (t) Romaine W/Shutterstock; (b) Mike Workman/Shutterstock; p36 AaronChenPS2/Shutterstock; p37 (t) Evgenii Panov/Shutterstock; (m) Murtaza.Ali/Shutterstock; (b) wk1003mike/Shutterstock; p38-39 (bg) Jenya Smyk/Shutterstock; p39 (tl) chanida pp/Shutterstock; (tm) New Africa/Shutterstock; (tr) rsooll/Shutterstock; (ml) amorehenrikson/Shutterstock; (mr) Mega Pixel/Shutterstock; (mr) Ter_Shutterstock/Shutterstock; (bl) Cobalt S-Elinoi/Shutterstock; (bm) Teerasak Ladnongkhun/Shutterstock; (br) Ad Oculos/Shutterstock; p40 Rich Carey/Shutterstock; p41 (t) Deemerwha studio/Shutterstock; (b) Prostock-studio/Shutterstock; p42 (b) Kabardins photo/Shutterstock; (b) swavo/Shutterstock; p43 (t) Multishooter/Shutterstock; (m) Derek Hatfield/Shutterstock; (b) Reshetnikov_art/Shutterstock; p44-45 (recycling boxes) Evan Lorne/Shutterstock; p44 (b) photka/Shutterstock; p46 (l) Ivan Chistyakov/Shutterstock; (m) Eric Isselee/Shutterstock; (r) Nynke van Holten/Shutterstock; p47 (bg) New Africa/Shutterstock; p48 Damsea/Shutterstock; p49 (t) WildMedia/Shutterstock; (b) BearFotos/Shutterstock; p50 Jahangir Jrs/Shutterstock; p51 (t) kukurund/Shutterstock; (b) New Africa/Shutterstock; p52 Susan Hodgson/Shutterstock; p53 (t) Danny Ye/Shutterstock; (b) Protasov AN/Shutterstock; p54 (t) ChiccoDodiFC/Shutterstock; (b) Dr Tushi Laxmi Suwal, used with permisson/; p55 (t) plavi011/Shutterstock; (b) Jonas Gruhlke/Shutterstock; p56 Alexandre.ROSA/Shutterstock; p57 (t) Rich Carey/Shutterstock; (b) Benny Marty/Shutterstock; p58 (bl) Marcio I. Sa/Shutterstock; (br) New Africa/Shutterstock; p59 (t) jindrich_pavelka/Shutterstock; (b) Charlotte Bleijenberg/Shutterstock; p60 (t) The Studio/Shutterstock; (b) yut548/Shutterstock; (br) yut548/Shutterstock; p61 Roman Samborskyi / Artsem Martysiuk/Shutterstock; p62 Asier Romero/Shutterstock; p63 (t) Iryna Inshyna/Shutterstock; (m) Yavdat/Shutterstock; (b) Laura Neate, used with permission; p64 (m) KatMoys/Shutterstock; (b) Krakenimages.com / InesBazdar/Shutterstock; p65 (t) Juice Flair/Shutterstock; (b) photomaster/Shutterstock; p66-67 (b) paula french/Shutterstock; p67 (t) elamie/Shutterstock; (m) clarst5/Shutterstock; p68 Leonardo Dantas Teixeira/Shutterstock; p69 Damsea/Shutterstock; p71 (t) Hazel Plater/Shutterstock; (b) Alexander Raths/Shutterstock; p72 (b) Pormezz/Shutterstock; p73 (t) Zoeytoja/Shutterstock; p74 Anggalih Prasetya/Shutterstock; p75 (t) Timolina/Shutterstock; (b) Fortyforks/Shutterstock; p76 (t) Kiian Oksana/Shutterstock; p76 (t) ppart/Shutterstock; (m) Pixel-Shot/Shutterstock; (b) Den Rozhnovsky/Shutterstock; p77 (t) goir/Shutterstock; (m) New Africa/Shutterstock; (b) Kitch Bain/Shutterstock; p78 oasisamuel/Shutterstock; p79 (t) Fabrizio Andrea Bertani/Shutterstock; (b) Refox Photos/Shutterstock; p80 YJPTO/Shutterstock; p81 (t) kckate16/Shutterstock/Shutterstock; (b) Prostock-studio/Shutterstock; p82 (t) LungLee/Shutterstock; (b) Cheryl E. Davis/Shutterstock; p83 (t) PARALAXIS/Shutterstock; (m) Mulad Images/Shutterstock; (b) Yellow Cat/Shutterstock; p84 (b) KK Tan/Shutterstock; p85 (t) ND700/Shutterstock; (b) New Africa/Shutterstock; p86 Ernest Rose/Shutterstock; p87 (t) Natali Kuzina/Shutterstock; (b) CGN089/Shutterstock; p88 (b) muratart/Shutterstock; p89 (t) Sydney Michalski/Shutterstock; (b) triocean/Shutterstock; p90 (bg) All for you friend/Shutterstock; p91 (t) otabalyuk/Shutterstock; (m) New Africa/Shutterstock; (b) Krakenimages.com/Shutterstock; p92 KatMoys/Shutterstock; p93 New Africa/Shutterstock; p94 Natalya_Maisheva/Shutterstock; p95 (t) Daria Yachmeneva/Shutterstock; (b) grublee/Shutterstock; p97 (icons) Pictogram studio/Shutterstock; p98 vesnushka/Shutterstock; p99 (t) A.Khachachart/Shutterstock; (b) Craig139/Shutterstock; p100 (t) Vidya Rahmandari/Shutterstock; (b) Liv Oeian/Shutterstock; p101 Susmit Das/Shutterstock; p102 cometa geo/Shutterstock; p103 Alones/Shutterstock; p104 (t) Scharfsinn/Shutterstock; (b) Dr Florence Gschwend, used with permission; p105 (t) Ian Dewar Photography/Shutterstock; (b) Spyros Vasileiou/Shutterstock; p106 (t) LeManna/Shutterstock; (b) Toa55/Shutterstock; p107 (t) 4045/Shutterstock; (b) John Gomez/Shutterstock; p109 (tl) New Africa/Shutterstock; (tr) grey_and/Shutterstock; (m) Sergiy Kuzmin/Shutterstock; p110 MEE KO DONG/Shutterstock; p111 Heidi Besen/Shutterstock; p112 Senyuk Mykola/Shutterstock; p113 RW Jemmett/Shutterstock; p114 Firn/Shutterstock; p115 (t) Colin Seddon/Shutterstock; (b) Anton Gvozdikov/Shutterstock; p116 trancedrumer/Shutterstock; p117 Wagner Meier / Stringer/Getty Images; Back cover (tr) HuHu/Shutterstock; (mr) Macrovector/Shutterstock; (bl) TWINS DESIGN STUDIO/Shutterstock.

INDEX

air pollution 36–37
albedo effect 29
algae 56–7, 119
Amazon rainforest 50–1, 58
Antarctic 28–9
Arctic 28
atmosphere 10–11, 119

biodiversity 6, 46
building and construction 112–13

carbon 14–16, 68, 79, 113, 119
carbon dioxide (CO_2) 12–13, 20, 27, 33, 46, 119
climate and climate change 2, 4–5, 8–26, 30–1, 68, 119
clothing and fashion 86–7, 90–1, 94–5
coal 16–17
compost 71, 119
conservation 33, 58–9, 119
cooking 76–7

decarbonisation 15
deforestation 12, 50–1, 117, 119

earthquakes 30
eco-tourism 98
electric cars 107
electricity 16–17, 21–3, 80–1
endangered species 47, 49, 54, 120

farming 68–9, 119, 121
flooding 9, 27
food waste 72–3
fossil fuels 16–17, 41, 96, 120
fracking 17, 120

glaciers 28
global warming 8–9, 11, 19, 27, 32–3, 66–7, 120
Great Barrier Reef 56–7
greenhouse effect 10–11, 120
greenhouse gases 1, 10–16, 19, 38, 68–9, 96–7, 120
greenwashing 114–15

health products 62–3

ice ages 2–3, 120
ice melt 9, 28
indigenous communities 34, 117, 120
Industrial Revolution 3, 10, 120

invasive species 52–3, 120

landfill 38–9, 62, 95, 120–1
laundry 92–3
leisure industry 78–9
light bulbs 21

mangroves 48, 121
marine conservation 35, 55–7, 59, 119
methane (CH_4) 13
microplastics 1, 86, 121
mining 17, 48, 121

natural disasters 30–1, 34
net zero 19, 121
nuclear power 25

oxygen 16, 46, 121

packaging 62–3
palm oil 51
photosynthesis 46–7, 50
plant-based diets 74–5
plastics 1, 3, 6, 40–1, 44, 73, 84, 86
poaching 54, 121
pollution 3, 6, 36–7, 40–1, 64–5

radiation 11
radioactive waste 25
recycling 44–5, 85
regenerative farming 69, 121
renewable energy 16, 22–5, 27, 121
rewilding 49

saving energy 20–1, 76–7, 92
sea levels rising 28, 34
sewage 64–5
space tourism 102–3
sport 78–9

toy industry 82–3
transport and travel 96–107
trees and forests 15, 50–1, 67, 79, 83
tsunamis 30

vegetarian and vegan diets 74–75
volcanos 31

waste 25, 38–45, 62–5, 72–3, 84, 95, 105, 120–1
water 12, 121
weather 8–9, 26–35, 66–7, 119
wildfires 9, 27, 121

121